AF830588

Fine Particle Science and Technology Series

John Keith Beddow, Editor-in-Chief

Professor, Chemical and Materials Engineering
Division of Materials
University of Iowa
Iowa City, Iowa

Advanced Particulate Morphology

Editors

John Keith Beddow
Professor, Chemical and Materials Engineering
Division of Materials
University of Iowa
Iowa City, Iowa

T. P. Meloy
Benedum Professor
Minterals and Engineering
University of West Virginia
Morgantown, West Virginia

Future Volume Topics*

Physical-Chemical Properties of Fine Particles

Particle Size Analysis

Instrumentation for Fine Particle Characterization

Particle Packing

On-Line Instrumentation

Particle Formation and Production

Fluid-Particle Separation

Fluid-Particle Interactions

Mixing of Particulate Solids

Bulk Solids Handling and Storage

Hazards

Agglomeration

Education and Research

Minerals Science

* Subject to change.

Advanced Particulate Morphology

Editors

John Keith Beddow
Professor, Chemical and Materials Engineering
Division of Materials
University of Iowa
Iowa City, Iowa

T. P. Meloy
Benedum Professor
Minerals and Engineering
University of West Virginia
Morgantown, West Virginia

Fine Particle Science and Technology Series
John Keith Beddow, Ph.D.
Editor-in-Chief

Proceedings of the National Science Foundation Residential Workshop on Advanced Particulate Morphology, University of Iowa, Iowa City, August, 1977.
Grant Number ENG 77-00980.

CRC Press
Taylor & Francis Group
Boca Raton London New York

CRC Press is an imprint of the
Taylor & Francis Group, an **informa** business

First published 1980 by CRC Press
Taylor & Francis Group
6000 Broken Sound Parkway NW, Suite 300
Boca Raton, FL 33487-2742

Reissued 2018 by CRC Press

© 1980 by CRC Press, Inc.
CRC Press is an imprint of Taylor & Francis Group, an Informa business

No claim to original U.S. Government works

This book contains information obtained from authentic and highly regarded sources. Reasonable efforts have been made to publish reliable data and information, but the author and publisher cannot assume responsibility for the validity of all materials or the consequences of their use. The authors and publishers have attempted to trace the copyright holders of all material reproduced in this publication and apologize to copyright holders if permission to publish in this form has not been obtained. If any copyright material has not been acknowledged please write and let us know so we may rectify in any future reprint.

Except as permitted under U.S. Copyright Law, no part of this book may be reprinted, reproduced, transmitted, or utilized in any form by any electronic, mechanical, or other means, now known or hereafter invented, including photocopying, microfilming, and recording, or in any information vstorage or retrieval system, without written permission from the publishers.

For permission to photocopy or use material electronically from this work, please access www.copyright.com (http://www.copyright.com/) or contact the Copyright Clearance Center, Inc. (CCC), 222 Rosewood Drive, Danvers, MA 01923, 978-750-8400. CCC is a not-for-profit organization that provides licenses and registration for a variety of users. For organizations that have been granted a photocopy license by the CCC, a separate system of payment has been arranged.

Trademark Notice: Product or corporate names may be trademarks or registered trademarks, and are used only for identification and explanation without intent to infringe.

Library of Congress Cataloging-in-Publication Data

National Science Foundation Residential Workshop
on Advanced Particulate Morphology, University
of Iowa, 1977.
Advanced particulate morphology.

(Fine particle science and technology ; v. 1)
Bibliography: p.
Includes index.
1. Particles— Congresses. I. Beddow, John K.
II. Meloy, T. P. III. United States. National
Science Foundation. IV. Title. V. Series.
TP156.P3N37 1977 620'.43 79-14310
ISBN 0-8493-5781-0

A Library of Congress record exists under LC control number: 79014310

Publisher's Note
The publisher has gone to great lengths to ensure the quality of this reprint but points out that some imperfections in the original copies may be apparent.

Disclaimer
The publisher has made every effort to trace copyright holders and welcomes correspondence from those they have been unable to contact.

ISBN 13: 978-1-315-89044-9 (hbk)
ISBN 13: 978-1-351-06954-0 (ebk)

Visit the Taylor & Francis Web site at http://www.taylorandfrancis.com and the
CRC Press Web site at http://www.crcpress.com

PREFACE

This volume, based on the 1977 NSF Residential Advanced Particle Morphology Workshop, covers the principle developments in this new and rapidly evolving field.

Chapter 1 is a review of particle morphology analysis. In it is described how the particle shape may be digitized and transformed into a set of Fourier (or Walsh) coefficients. These coefficients in themselves have physical significance (for example, A_3 is an indicator of triangularity), and they may also be transformed into other morphology descriptors. The sets of coefficients corresponding to the shape of the particle profile constitute information which may then be further processed through classifiers; or the data may be treated in the form of a signature; or the data may be subject to harmonic analysis.

Chapter 2 is a discussion of how the coefficients can be summarized so as to give an economical, position invariant, universal signature.

Chapter 3 describes some very interesting geological examples in which particle morphology analysis has been used very successfully. An important aspect of this work is the systematic use of a set of statistical methods in harmonic analysis of particle shape Fourier coefficients.

Chapter 4 gives an elegant description of the use of fuzzy sets in particle shape analysis with specific reference to the Fuzzy ISODATA.

Chapter 5 is an in-depth look at some of the orthonormal spanning sets used to transform silhouette data into coefficients. Walsh, Fourier, Haar, and the optimum Karhunen-Loeve transforms are reviewed as to the applicability of the particle analysis.

When particle shapes are analyzed, the analysis may be viewed as part of the general field of pattern recognition. Chapter 6 approaches particle morphology analysis from this point of view and gives a general outline of image processing and recognition.

Particles are not isolated from one another, and particle sets or assemblies have structure. Chapter 7 deals with an analysis of the randomly packed structure of particles. The approach is based upon certain general similarities between packing structures of molecular assemblages and fine particle sets.

The divergent fields of interest of the authors reflect the prospects for the wide application of particle morphology analysis and include: many branches of science and engineering concerned with fine particles, information processing, life sciences, pharmacy, and food technology.

JKB
TMP

THE EDITORS

John Keith Beddow received his Ph.D. in Metallurgy from Cambridge University, England, in 1959. Currently President of the Fine Particle Society, he is a member of the Faculty at the University of Iowa, where he heads a small research group in fine particle science with emphasis on morphological analysis. Dr. Beddow is an active lecturer and author. He has also been active as a Consultant in metallurgy, powder metallurgy, and powder technology for numerous corporations. His present research activities are in powder metallurgy and technology and particle morphological analysis. Dr. Beddow is married, with four daughters and has resided in the U.S. since 1966.

Thomas P. Meloy received his Ph.D. in Metallurgy from the Massachusetts Institute of Technology in 1960. Dr. Meloy has had a distinguished career in industry, government (where he was Director of the Engineering Division of the National Science Foundation), and in academia. Currently, he is Benedum Professor of Mineral Processing at the University of West Virginia where he serves as department chairman. He leads a research group in morphological analysis of particulates in mineral processing. Dr. Meloy is active in public service and committee work. He is the founder and Editor of the *Journal of Ocean Technology*. His present research interests are particle characterization, effect of particle shape on the behavior of particle systems, analysis of beneficiation circuits, the optimization of beneficiation circuits and theory of particle separation.

CONTRIBUTORS

John Keith Beddow
Professor
Department of Chemical and Materials Engineering
University of Iowa
Iowa City, Iowa

James C. Bezdek
Associate Professor
Department of Mathematics
Utah State University
Logan, Utah

Jeffrey Brown
Geology Department
University of South Carolina
Columbia, South Carolina

Robert Ehrlich
Professor
Geology Department
University of South Carolina
Columbia, South Carolina

Duane T. Eppler
Geology Department
University of South Carolina
Columbia, South Carolina

Keishi Gotoh
Professor of Chemical Engineering
Department of Fisheries Chemistry
Hokkaido University
Hakodate
Japan

Thomas P. Meloy
Benedum Professor
West Virginia University
Morgantown, West Virginia

G. Robert Redinbo
Associate Professor
Electrical and Systems Engineering Department
Rensselaer Polytechnic Institute
Troy, New York

Julius T. Tou
Graduate Research Professor and Director
Center for Information Research
University of Florida
Gainesville, Florida

Jeffrey M. Yarus
Geology Department
University of South Carolina
Columbia, South Carolina

ACKNOWLEDGMENT

The work reported in this book is one result of a Residential Research Workshop in Advanced Particulate Morphology conducted at the University of Iowa by one of the editors (J. K. Beddow) in August, 1977. Grateful thanks to the Particulates Processing and Multi-Phase Flow Program of the National Science Foundation and the program director, Morris Ojalvo, is hereby acknowledged.

TABLE OF CONTENTS

ALPHABETICAL SYMBOLS

a_n	Coefficient
A	Non-fuzzy subset
A	Area
A_0	Zero Fourier coefficient or Walsh coefficient (mean radius)
A_1	First Fourier or Walsh coefficient (error in locating centroid)
A_2	Second Fourier or Walsh coefficient (elongation)
A_3	Third Fourier or Walsh coefficient (triangularity)
A_4	Fourth Fourier or Walsh coefficient (squareness)
A_i	Fourier i^{th} coefficient
A_n	Fourier coefficient
A_I	Intercept on signature curve
A_K	Fourier coefficient
A_R	Aspect ratio
b_n	Coefficient
b_N	A constant
B	Breadth
B	Bandwidth of waveform
B	{all cases of beer}
B_k	Fourier Coefficient
BANKR↑	Bankruptcies increase
c	Number of fuzzy subsets
c	Cutoff point
c_1, c_2	Two plain beer cartons
c_i^*	Undetermined constants
$c_{k,r}$	Spectrum for Fourier series representation of correlation
$c_{k,z}$	Spectrum for Fourier series representation of convolution
$\lvert c_{k,z}\rvert^2$	Power spectrum
c_n	Coefficients
$C_s(k)$	Transform coefficients of the shift sequence X(((M + S)))
$C_{xy}^{(\tau)}$	Convolution between two waveforms
$C_{xy}^{(\omega)}$	Fourier transform of $C_{xy}^{(t)}$
d	Molecular diameter
d	Number of features per observation
d(x)	Decision function
dR	Thickness of spherical shell
D(k)	A summation
DCT	Discrete cosine transform
DFT	Discrete Fourier transform
DJ↓	Dow Jones Index decline
$D_L(z)$	Discriminant function
$D\ell(z)$	Discrimination function
DWT	Discrete Walsh transform
E(k)	A summation
$f_A(x)$	Membership function of x in A
$f_{NO}(t)$	Optimum linear approximation
$f_p(t)$	Periodic extension for a specific interval
f(R)dR	Probability that center to center distance between particle and its nearest neighbor lies between R and R + dR from the given central particle
f_s	Sampling frequency
f(t)	A function
F	Finite fuzzy subset
F_c	Partition coefficient
FFT	Fast Fourier transform
$F_L(\omega)$	Fourier transform of signal duration L
F_v	Horizontal Feret diameter
$F(\omega)$	The forward Fourier transform
FWT	Fast Walsh transform
g	Radial distribution function
g_k	A grain
G	Fuzzy subset
GNP↓	Gross national product decline
h	Planck constant

ALPHABETICAL SYMBOLS (Continued)

h(p)	Height
har(r,m,t)	Haar function
H_c	Entropy
i	Subscript, one of many
I	An interval
I	NxN identity matrix
I_h	Number of horizontal intercept
I_v	Number of vertical intercept
j	Arc length of jth vertex from starting point
j	Kind j — type j species
$J_f(u,v)$	A functional, within group sum of squared errors
J_m	Generalized total square error
$J_m(u,v)$	A generalized within group sum of squared errors functional
J_{wx}	A criterion function
k	An integer, i.e., 0,1,2,3,....
k	A coefficient
k	Boltzman's constant
k	0,1,2,...N
K	A constant
KLT	Karhunen-Loève transform
ℓ	Length of perimeter
L	Arc length
L	Length
L	Lumpiness
L_3	Mean particle interception three dimensions
LDF Rule	Linear discriminant function rule
$L_{(k)}$	Coefficients for DCT
m	$1,2,\ldots 2^r$
m	$0,1,2,\ldots n-1$
m	An integer
m	A coefficient
m	Molecular mass
m	A power
M_c	Set of all nondegenerate comparators of x
M_{fc}	Set of all nondegenerate fuzzy c-partitions of x
max	Maximum
Min	Minimum
MIN	Minimum
n	An integer
n	Number of features
n	Order of coefficient
N	A finite number
N	The cordinality of the basis set
N_v	Number of density of particles
p	A person
p	Pressure
Pr	Posteriori probability
PL	Piecewise linear function
r	A person
r	0,1,....
r	Mean radius
$r_{xy}(\tau)$	Periodic correlation function
R	Radius vector
R	Roughness
R	A distance in units of sphere diameter, radius of spherical shell
Re	Real
$R_{xy}(\omega)$	Transform domain
R transform	A term of the FWT ($R_x(z)$ rank of $D_x(z)$)
s	Slope of signature curve
s	A shift
SF1, SF2, SF3, SF4	Shape factors
S_n	An exponent
S_n	$n = 1,2,3,\ldots$ a set
$S(u,\alpha,\beta,\gamma)$	Fuzzy membership function
t	Interval $\{0,2\pi\}$
t_0	Starting position
T	Length
T	A matrix
T	Absolute temperature
T	Thickness
u_i	Characteristic function
u_{ik}	Elements of U
u_n	Members of A
U	Fuzzy subset
U	A matrix

ALPHABETICAL SYMBOLS (Continued)

U(p)	Characteristic function of people
U_r	First half of data sequence
UNEMP	Unemployed
v	Cluster center
v	$(v_1, v_2, \ldots\ldots v_i)$ prototype
v_f	Free volume of a single molecule
V	Total volume
V_{cn}	Vector space of all real (cxn) matrices
V_e	Volume of liquid phase
V_f	Empty volume of the system
V_r	Second half of data sequence
w	Pitzer's acentric factor
w_i	Weighting parameter
w	Augmented weight vector
W	The nth root of unity
W(n,t)	Walsh function
x	Feature vector
x_0	A constant
x_1	General coordinate variable
x(t)	Original waveform
$x(\omega)$	Fourier transform of x(t)
X	A vector
$X_c\mathbf{R}^d$an	n Sample from $\mathbf{R}^d$
X_keX	A feature vector in X
$X_{k,j\varepsilon}$	Measured features of X_k
Y	A vector
$Y(\omega)$	Fourier transform of y(t)
Z	Coordinate number
Z	An observation
$Z_i \ldots Z_k$	Nearest points to Z by a distance measure
$Z_{xy}(\tau)$	Periodic convolution function
Z_m	A convolution

GREEK SYMBOLS

$\propto$	A real number
$\propto$	Integer ranging from 0 to 3
$\propto_i$	i^{th} Phase angle Fourier transform
$\propto_n$	A constant
$\propto_n$	Phase angle
β	Integer ranging from 0 to 3
$\{\beta_k\}_{k=0}^{N-1}$	Any set of constants
$\{\gamma_k\}$	Optimum coefficients
Γ_n	A constant
δ_{ij}	Kronocker delta function
Δ	Difference, interval, part of
ε	Tolerance of measuring device
$(-\varepsilon)$	Potential energy
$\varepsilon(t)$	Error of approximation
$\varepsilon_m{}^2$	Mean square average of ΔX
$\|\|\varepsilon^2\|\|$	Mean square error criterion
θ	Angle
λ	Molecular shape parameter
λ	A given translation
λ	Complex numbers
μ_O	First Fourier Coefficient, (ϕ,ℓ) method
μ_i/u_i	Grade membership of u_i in F
μ_1,μ_2	Normally distributed means
$\pi(u,\beta,\gamma)$	Fuzzy membership function
π_1,π_2	Populations
ϱ	Liquid density
$\varrho,\varrho_E,\varrho_R$	Constants
Σ_x	Covariance of vector x
Σ	Covariance matrix
Φ	Bulk mean volume fraction of particles
Φ_b	Boiling mean volume fraction of particles
Φ_c	Critical mean volume fraction of particles
$\Phi_i(x)$	Linearly independent bounded function of x in Σ^n
$\Phi(\ell)$	Change of shape
Φ_m	Melting mean volume fraction of particles
Φ_n	A set of functions
Φ_N	Vectors (rows in orthogonal matrix T)
$\Phi^*(t)$	A function
Ψ	Angle of pyramid apex (compliment $\propto$)
Ψ	Potential energy between one molecule and all others
ω	$2\pi/T$ fundamental parameter

MATHEMATICAL SYMBOLS

	Transpose	$\|\,\|$	Absolute value
∪	Union	((x))	Modulo N operations
∩	Intersection	sal	Composite term from sin/Walsh
○	Dyadic operation		
○	Bounded sum	cal	Composite term from cosine/Walsh
○	Ring sum		
○	Bounded difference	Wal	Walsh
∨	Maximum	WAL	Walsh
∧	Minimum	PL/1	Computer language
Δ	Defined as	RADNO	A computer program title
∫	Union of fuzzy singletons μ_i/u_i	$\underset{\to}{x}t$	Average of x
u		x	Transpose of x
F′	Compliment of F	ε	Belongs to
X x Y	Cartesian product	*	Complex conjugation
R o S	Composition of R and S	R^d	Feature space
{ }	A set		

Chapter 1

PARTICLE MORPHOLOGICAL ANALYSIS

J. K. Beddow

TABLE OF CONTENTS

I. SIGNIFICANCE OF MORPHOLOGICAL ANALYSIS — AN ENGINEERING DISCIPLINE

Historically, particle size has been defined by an interaction with a measuring instrument such as a sieve; and particle shape, by one or two parameters such as size and aspect ratio. Recently particle silhouettes have been analyzed by sophisticated signal processing techniques such as Walsh and Fourier series, yielding numerical values which describe the particle profile shape with great accuracy. Using existing laboratory equipment it is possible to obtain particle fingerprints with which the particle can be uniquely identified. Corresponding sets of data can be obtained for powders in the form of signature analysis or harmonic analysis. These data can be used to determine whether a sample is composed of powder A or powder B, is a combination of A and B, or if A and/or B is present in powder sample. For example, if two power-plant stacks emit particulates, it is possible to determine at a sampling station how much of the particle sample was contributed by the first stack and how much by the second. By using the advanced clustering and classification technique, the number of particle shapes is determined. For the homogeneous powders we have examined thus far, all have multimodal shape characteristics. This unexpected result has deep theoretical implications both for morphological and statistical studies.

Application of the morphological analysis techniques has widespread applications in industry for powder specification and process control, in government for forensic and environmental problems, and in medicine for the identification of pathogenic particles.

II. FOURIER, WALSH, AND HAAR FUNCTIONS

A. The Method of Fourier Analysis of a Profile

It has been shown that a particle profile can be considered a population of points from which a sample set of points in the form of (x,y) coordinates can be abstracted.[1,2] This sample set must be a Nyquist set, i.e., the points must be evenly spaced round the profile and the software has to be adjusted to do this. Using the (x,y) data, the centroid of the shape is calculated and thereafter the set of coordinates is transformed into polar coordinates, (R,Θ). The function (R,Θ) is then described in terms of a Fourier series of the general form:

$$R(\theta) = A_0 + \sum_{n=1}^{\infty} A_n \cos(n\theta - \alpha n) \quad (1)$$

in which A_n are the Fourier coefficients and αn are the phase angles. It may be noted also that in terms of the more standard form of the Fourier relationship:

$$A_n = \left(a_n^2 + b_n^2 \right)^{1/2} \quad (2)$$

$$\alpha n = \tan^{-1} b_n/a_n \quad (3)$$

in which a_n and b_n are coefficients such that:

$$R(\theta) = A_0 + \sum_{n=1}^{\infty} (A_n \cos n\theta + b_n \sin n\theta) \quad (4)$$

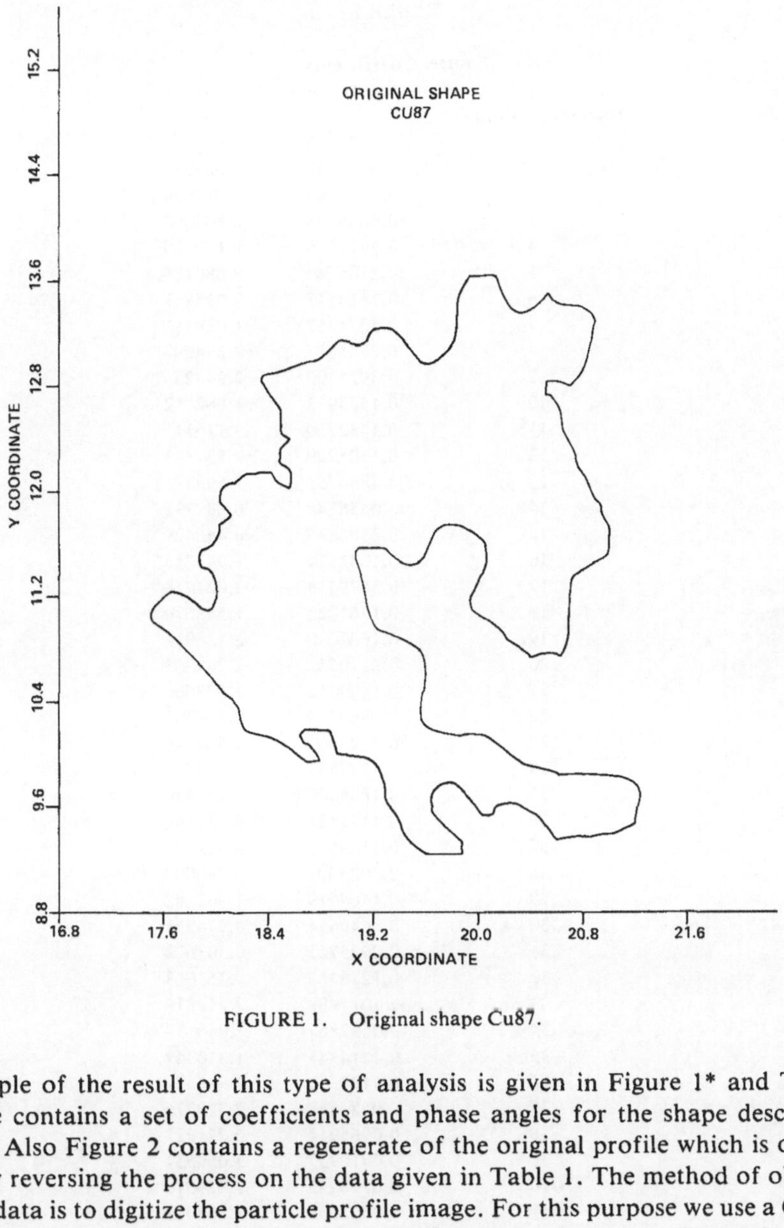

FIGURE 1. Original shape Cu87.

An example of the result of this type of analysis is given in Figure 1* and Table 1. The table contains a set of coefficients and phase angles for the shape described in Figure 1. Also Figure 2 contains a regenerate of the original profile which is obtained by simply reversing the process on the data given in Table 1. The method of obtaining the (x,y) data is to digitize the particle profile image. For this purpose we use a Grafpen system (Figure 3). When using this system, the operator places a profile of the subject particle on the Grafpen table and with the stylus hand-traces the profile. The tip of the profile emits a noise as it touches the surface being traced and this noise is picked up on sensitive microphones. These microphones run in a continuous bank along one side and top of the tabletop. In this way the (x,y) coordinates can be obtained. It is possible to adjust the rate at which data is acquired and so the size of the point sample obtained may be adjusted for each requirement. Errors associated with digitizing procedures have been fully discussed elsewhere.[3] Two points should be borne in mind. With a method in which the profile is traced, the human error is always present. This can be significantly reduced in the case of a system such as the Quantimet® with a FIFI attachment. Secondly, the error associated with the Fourier method can always be reduced by taking another term in the series.

* The (Φ, ℓ) method rather than the $(R\Theta)$ method was used here. See Section III.B.

TABLE 1

Fourier Coefficients

Harmonic Number	A_n	α_n
1	0.9180837	−0.026830
2	0.7811343	−1.919139
3	0.6038185	2.204062
4	0.5977379	1.177760
5	0.2106746	−0.880125
6	0.1914537	3.035851
7	0.5071257	−1.026711
8	0.2885617	−2.234097
9	0.3928300	−2.244235
10	0.1503933	−1.662912
11	0.1452250	3.116115
12	0.1505209	−2.837341
13	0.4840755	−1.598395
14	0.0456248	0.882792
15	0.2388835	−0.446429
16	0.1031974	1.943743
17	0.3679479	−3.021418
18	0.1361023	1.582670
19	0.1662030	2.128934
20	0.2110651	−2.249763
21	0.1499012	1.573666
22	0.0867854	2.674203
23	0.2423907	2.593346
24	0.1337547	1.733434
25	0.1888060	−1.773540
26	0.1171121	−0.422164
27	0.1184515	0.368735
28	0.2025601	0.740007
29	0.1464919	−1.462102
30	0.1449654	−0.974561
31	0.0925761	0.503672
32	0.1278262	0.361063
33	0.1011988	2.310310
34	0.0692867	2.040118
35	0.1214545	1.330347
36	0.1129320	2.927621
37	0.0784888	2.829776
38	0.0965914	0.771211
39	0.0910937	3.069534
40	0.0778255	−1.429016
41	0.0613601	−1.262012
42	0.0825907	−0.342230
43	0.0328851	2.987070
44	0.0710672	−0.856483
45	0.0336690	2.626568
46	0.0656527	−0.647202
47	0.0577465	1.744238
48	0.0996903	−2.300911
49	0.0355991	−1.061722
50	0.0286593	−1.726463

One great advantage of the Fourier method used in this way is that the individual coefficients have real significance in terms of the shape of the profile. For example, plots of $A_0 + A_1 \cos\Theta$ versus Θ; and of $A_0 + A_2 \cos 2\Theta$ versus Θ; and of $A_0 + A_3 \cos 3\Theta$ versus Θ; and so on, are shown in Figure 4. The first diagram shows that A_0

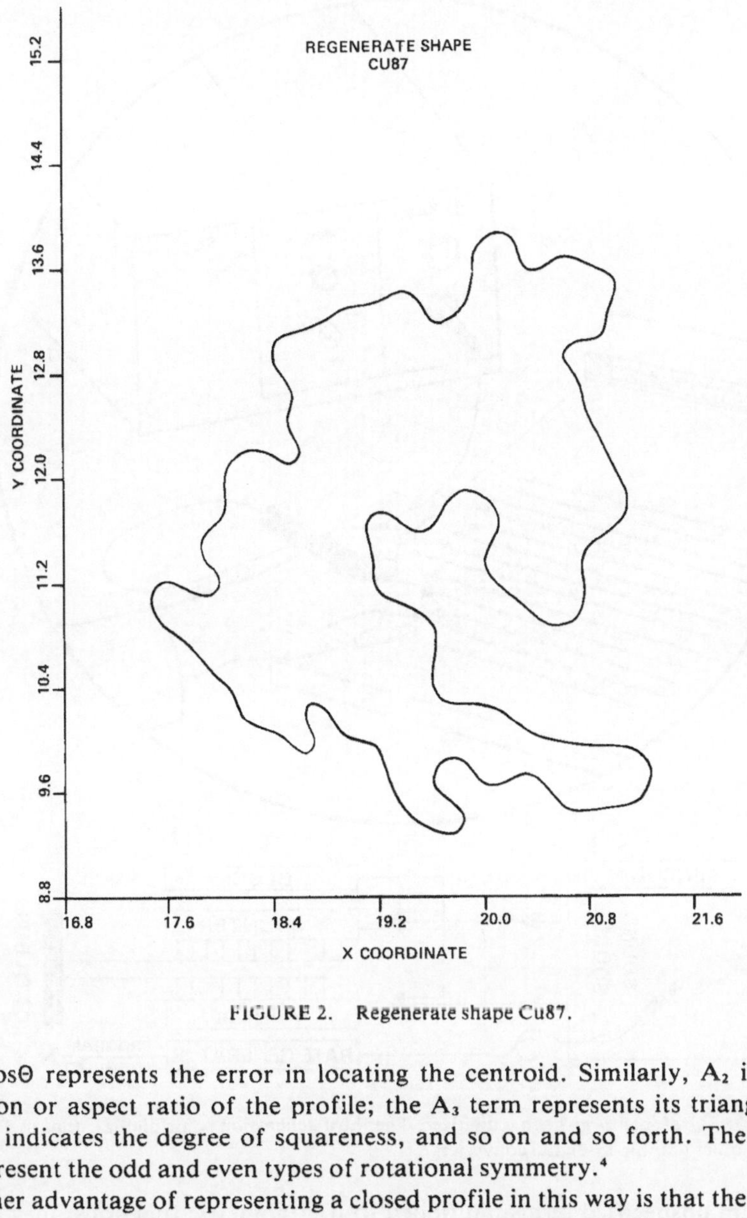

FIGURE 2. Regenerate shape Cu87.

+ $A_1 \cos\Theta$ represents the error in locating the centroid. Similarly, A_2 indicates the elongation or aspect ratio of the profile; the A_3 term represents its triangularity; the A_4 term indicates the degree of squareness, and so on and so forth. The coefficients thus represent the odd and even types of rotational symmetry.[4]

Another advantage of representing a closed profile in this way is that the successively higher-order coefficients decrease so that one can represent the (A_n) as a signature plot of A_n vs. n. This is referred to in Section VB. What this means is that the higher the order of the coefficient, the finer the morphological detail it represents.[5]

B. The Reasons for Using Walsh Functions[5]

Fourier functions operate over an interval 0 through 2π. They are orthogonal, normalized, and complete. The term orthogonal means that the product of any two functions integrated over an interval is zero, unless they are identical. The term normalized means that the integral of the function squared over the interval equals one. The term complete means (among other things) that all of the functions in the set are defined. However, there are potential advantages in using square-wave functions instead of the more traditional Fourier and for a number of reasons. For example, it is possible to state the Fourier form of a line (or a surface) integral but unfortunately because the

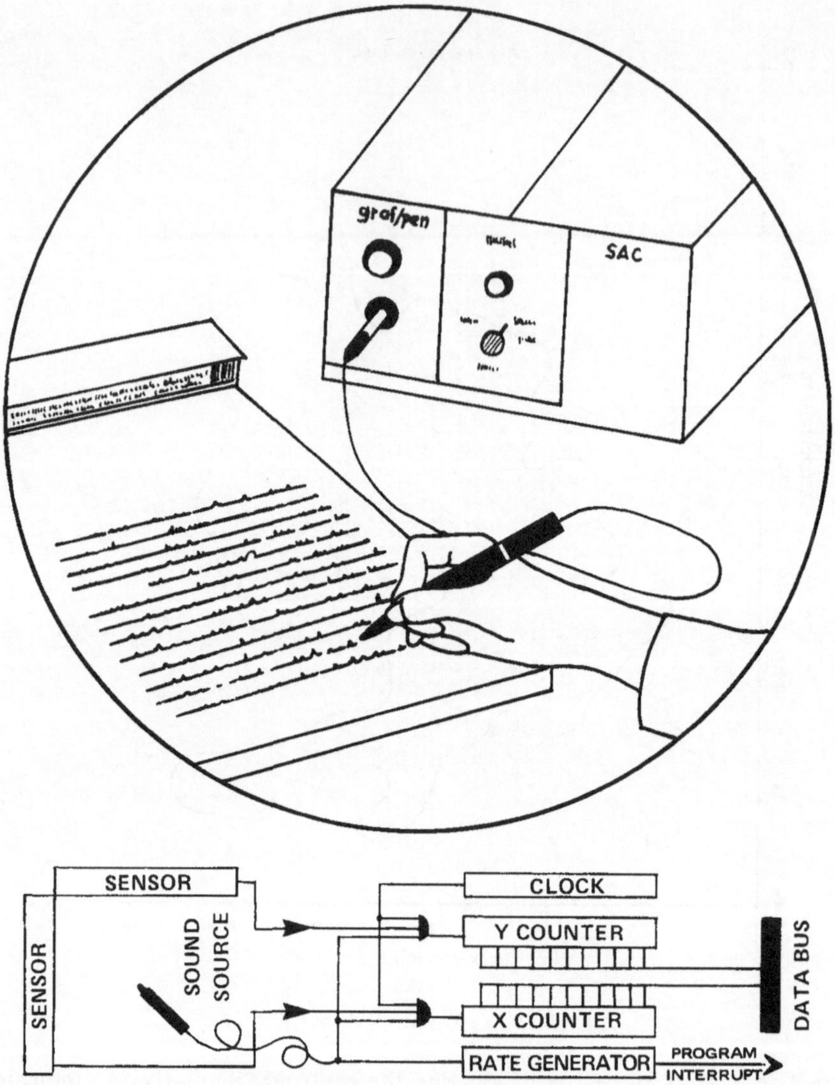

FIGURE 3. Graf/pen® gp3 sonic digitizer. The third generation sonic digitizer that obsoletes all other graphic-to-digital converters.

integrals contain unspecified terms multiplied to trigonometric functions, they cannot be integrated in closed form and so for this reason uses for Fourier representation of surface features appear to be limited. Alternatively, Walsh functions (which are square-waved functions as shown in Figure 6 of Chapter 2) can be used to represent the particle profile in just the same way that the Fourier method can be used. This is illustrated in Figure 5[8] where it can be seen that the Fourier and Walsh representations are equivalent. Furthermore, the individual coefficients have similar significance. As shown in Figure 4, Chapter 2, A_2 in the Walsh representation indicates elongation or aspect ratio just as does the Fourier A_2. Similarly, in Figure 5, Chapter 2, A_3 indicates triangularity and so on. One refers to Figures 6 and 7 to begin to see the advantages of the Walsh representation. Thus, the excess profile can be modeled as the total length of the vertical edges and this is 8 A_2 in Figure 7. The advantages of doing this sort of arithmatic compared to the problems associated with trying to calculate surface features using the Fourier series are obvious.

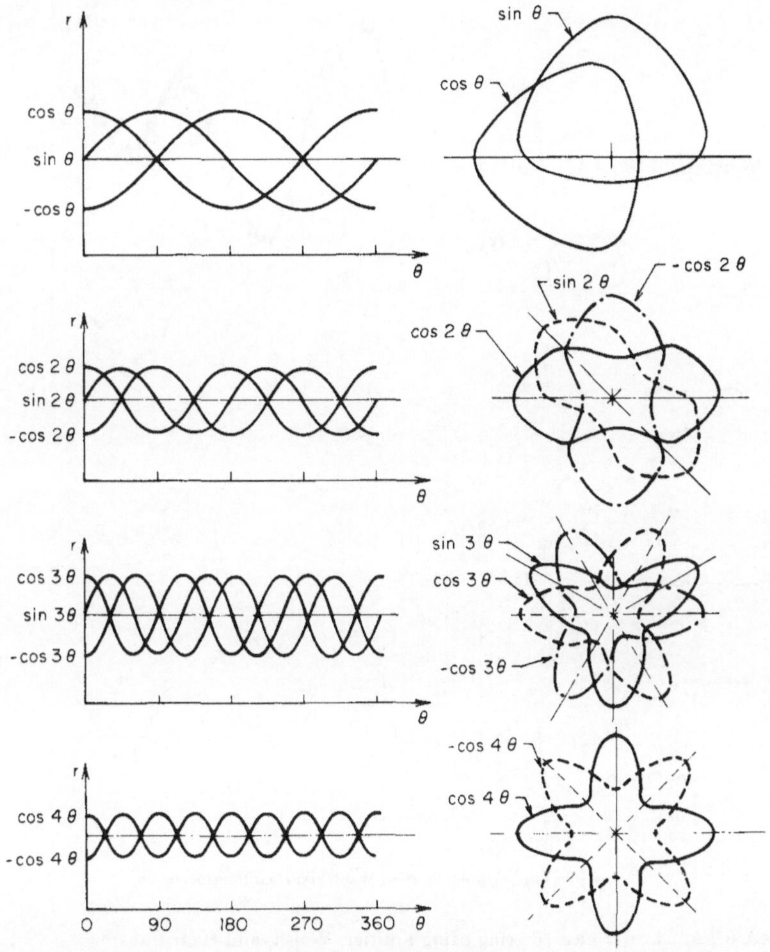

FIGURE 4. Plots of regenerates for various As. (From Beddow, J. K., Philip, G. C., Nasta, M. D., and Vetter, A. F., *Powder Technol.*, 18, 19, 1977.

Referring again to Figure 6, Chapter 2, the term Walsh function was named in honor of the inventor of the function. The terms sal and cal are composite terms made up from sine/Walsh and cosine/Walsh, respectively. The single function wal(j.Θ) can be defined in order to relate the three functions:

$$\mathrm{Wal}(2i,\theta) = \mathrm{Cal}(i,\theta)$$
$$\mathrm{Wal}(2i-1,\theta) = \mathrm{Sal}(i,\theta) \qquad (5)$$

in which i is termed the sequency and is one half of the number of zero crossings per second. Note from examination of the Walsh displays in Figure 6, Chapter 2 that the intervals are not equally spaced as are those of the Fourier function. Note also that wal(o,Θ) is the same as R_0,; sal (2,Θ) is the same as R_2 and so on. (The Rs are Radamacher functions shown in Figure 8).

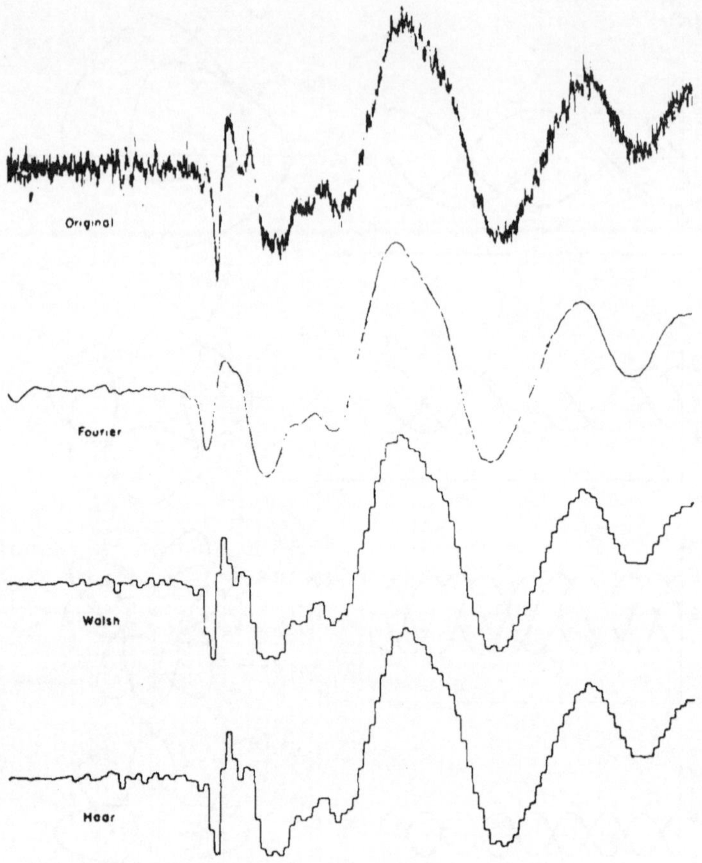

Low-pass filtering using Fourier, Walsh and Haar transformation.

FIGURE 5. Lower pass filtering using Fourier, Walsh, and Haar transformations. (From Beauchamp, K. G., *Walsh Functions and their Applications*, Academic Press, London, 1975. With permission.)

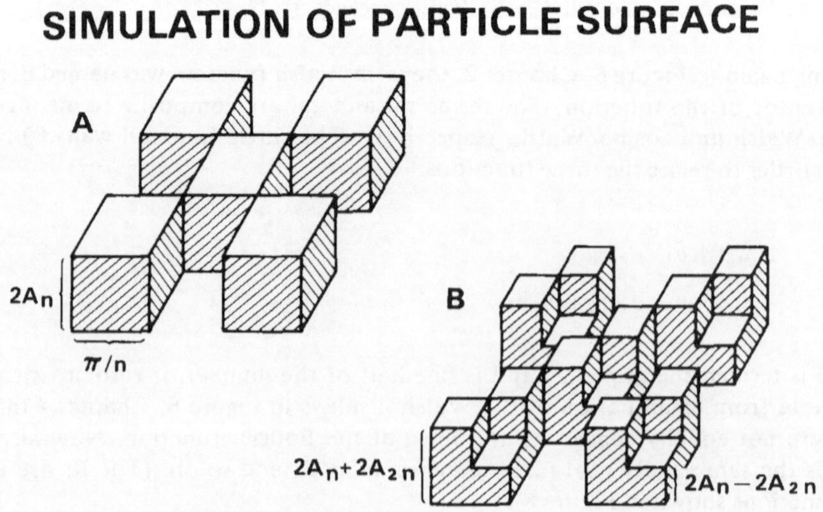

FIGURE 6. Simulation of particle surface. (From Meloy, T. P., *Powder Technol.*, 16(2), 233, 1977. With permission.)

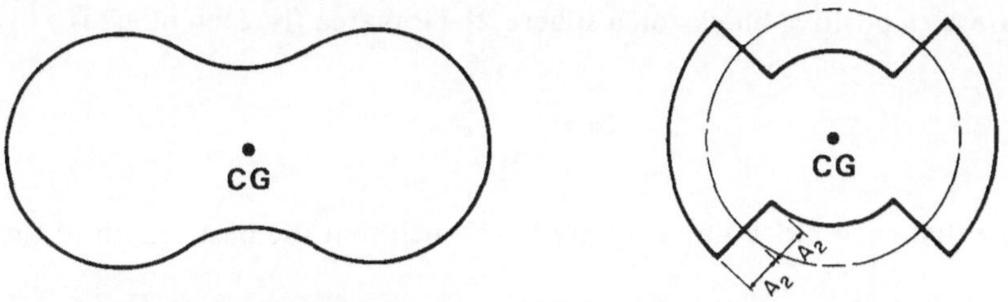

FIGURE 7. Illustrating the concept of excess edge length and its measurement.

1. Arc Length

With the treatment of Walsh functions in the line integral, the arc length can be described by two terms:

$$L = 2\pi A_0 + 8A_2 + 12A_3 + \ldots$$

$$= 2\pi A_0 + 4 \sum_{n=2}^{\infty} N \cdot An \qquad (6)$$

This equation seems to be reasonable, for example, if we draw out the superposition of A_2 and A_3, in which we shifted the peaks o A_3 with a 15° clockwise rotation in relation to those of A_2. But, if we shift 30°, the second term of excess edge length is reduced into:

$$8A_2 + 8A_3 + \ldots$$

Similar to the result from the 30° shift, we find that shifts of 60°, 90°, 120°, 150°, 10° 210°, 240°, 270°, 300°, 330°, and 360° also have the same expression. This means that Equation 6 is a translation variant.

C. Physical Significance of 3D Walsh Function

1. Cubing the Sphere

It is assumed that the number of square-wave peaks are in the ratio of $1/\pi$ between the flat square surface and the sphere surface. If the circumference of the sphere has n positive waves then there are n^2 positive waves on the flat square surface or n^2/π on the sphere surface.

The surface area of a sphere having a unit in radius is

$$A = 4\pi R^2 = 4\pi \qquad (R = 1)$$

By assumption of square waves, the total top area of positive blocks is

$$A/2 = 2\pi$$

There are n^2/π positive blocks on a sphere, the top area for each block is

$$2\pi/n^2/\pi = 2\pi^2/n^2$$

We assume that each block has a square base, such that the base length of the positive block is

$$\sqrt{2\pi^2/n^2} = \sqrt{2}\pi/n \qquad (7)$$

Each positive block on the sphere's surface has a height of An and a square base $\sqrt{2}\pi/n$ on a side.

2. Excess Surface Area

There are n^2/π of blocks each with four sides, a height of 2An, and a base length of $\sqrt{2}\pi/n$. Since only half of the side area is attributable to any one block, the side area for each block is

$$\tfrac{1}{2}(2An \times \sqrt{2}\pi/n \times 4) = 4\sqrt{2}\pi An/n$$

in which

$$2x \quad \frac{4\sqrt{2}\pi A_2}{2} = 4\sqrt{2}\pi A_2$$

$$4x \quad \frac{4\sqrt{2}\pi A_3}{3} = \frac{16\sqrt{2}}{3}\pi A_3$$

$$6x \quad \frac{4\sqrt{2}\pi A_4}{4} = 6\sqrt{2}\pi A_4$$

$$\frac{n^2}{\pi} x \quad \frac{4\sqrt{2}\pi An}{n} = 4\sqrt{2}nAn$$

The total side area of all blocks of all frequencies is

$$\text{Side Area} = 4\sqrt{2}\pi A_2 + \frac{16\sqrt{2}\pi}{3} A_3 + 6\sqrt{2}\pi A_4 + 4\sqrt{2}\sum_{n=5}^{\infty} nAn \qquad (8)$$

3. Edge Length

Each block has four vertical edge lines. Only one-fourth of them can be assigned to a block. The edge length for each block is

$$\tfrac{1}{4} \times 4 \times 2\,An = 2An$$

in which

$$2x(2A_2) = 4A_2$$

$$4x(2A_3) = 8A_3$$

$$6x(2A_4) = 12A_4$$

$$\frac{n^2}{\pi} x(2An) = \frac{2}{\pi} n^2 An$$

The resulting edge length is

$$\text{Vertical Edge Length} = 4A_2 + 8A_3 + 12A_4 + \frac{2}{\pi} \sum_{n=5}^{\infty} n \qquad (9)$$

The horizontal edge length on the top of ech block is

$$4X\left(\frac{\sqrt{2\pi}}{n}\right) = \frac{4\sqrt{2\pi}}{n}$$

and the total horizontal edge length is

$$\text{Horizontal edge length} = 4\sqrt{2\pi} + \frac{16\sqrt{2}}{3}\pi + 6\sqrt{2\pi} + 4\sqrt{2} \sum_{n=5}^{\infty} \qquad (10)$$

Meloy[5] proposed that a physical property of a particle can be represented by:

$$\text{Physical property } \alpha A_0 + K \sum_{2}^{\infty} n^{\gamma} A_n^{\beta} \qquad (11)$$

in which A_o = average radius of the particle, K = constant, α, β = integers 0→3, n = the number of the coefficient, and A_n = the value of the coefficient. Using this approach the set of equations shown in Table 2 has been developed.[5]

The first row of functions in Table 2 do not appear to represent a physically meaningful property except for A_0, which is the mean particle diameter. It should be noted that this mode of definition of mean particle diameter is rigorous and it is uniquivocal. Again in Table 2, we have already calculated the form of, for example, the edge length of Equation 9, that appears in row 2, column 3, the side area of Equation 8 which is in row 2, column 2. A 3D simulation of the particle surface is shown in Figure 8. This will help us to visualize the volume o the positive blocks and the top surface area.

$$\frac{\pi}{2} A_2 + \frac{8\pi^2}{9} A_3 + \frac{3\pi^2}{4} A_4 + 2\pi \sum_{n=5}^{\infty} A_n \qquad (12)$$

4. Volume of Positive Blocks

This term is found in Table 2, (row 2, column 1).

TABLE 2

The Meloy Equations
For Morphological Characteristics

β $\alpha\rightarrow$	0	1	2	3
0	A_o	$A_o+k_{21}\sum_2^{\infty} n$	$A_o+k_{31}\sum_2^{\infty} n^2$	$A_o+k_{41}\sum_2^{\infty} n^3$
1	$A_o+k_{12}\sum_2^{\infty} A_n$	$A_o+k_{22}\sum_2^{\infty} nA_n$	$A_o+k_{32}\sum_2^{\infty} n^2A_n$	$A_o+k_{42}\sum_2^{\infty} n^3A_n$
2	$A_o+k_{13}\sum_2^{\infty} A_n^2$	$A_o+k_{23}\sum_2^{\infty} nA_n^2$	$A_o+k_{33}\sum_2^{\infty} n^2A_n^2$	$A_o+k_{43}\sum_2^{\infty} n^3A_n^2$
3	$A_o+k_{14}\sum_2^{\infty} A_n^2$	$A_o+k_{24}\sum_2^{\infty} nA_n^3$	$A_o+k_{34}\sum_2^{\infty} n^2A_n^3$	$A_o+k_{44}\sum_2^{\infty} n^3A_n^3$

From Meloy, T. P., *Powder Technol.*, 16(2),233, 1977. With permission.

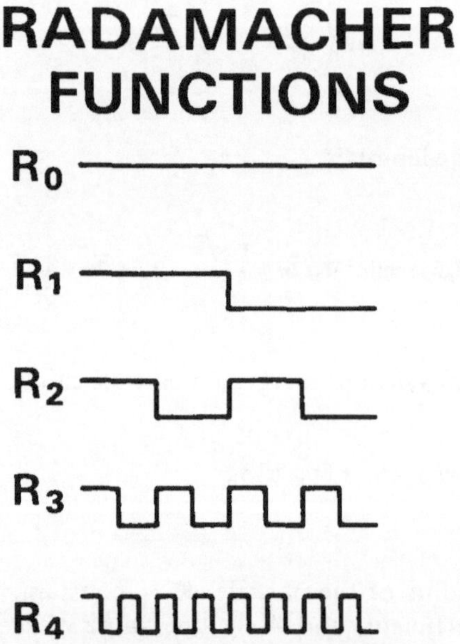

FIGURE 8. A set of Radamacher functions. (From Beauchamp, K. G., *Walsh Functions and their Applications,* Academic Press, London, 1975. With permission.)

5. Top Surface Area

In Table 2 (row 3, column 1) the terms may be predicted to represent excess top surface area. As we know, the total top surface area of all positive blocks is

$$\tfrac{1}{2}\times(4\pi A_0^2) = 2\pi A_0^2$$

in which $\beta = 2$. The equation is similar to that in Table 2 (row 3, column 1).

6. Activity Corners

In the application of the square waves to the physical properties of particles, $\Sigma n \propto An^{\beta}$, one must consider the relationship between n and An, and the determination of n when it approaches infinity. There is a linear relationship between log An and n, and the value of An closes to zero as n nears infinity (the signature). One can choose an upper limit of n from a meaningful physical parameter such as the unit crystal size — 10A, or choose a limit of n to truncate the formula without the loss of significance of the equation.

To calculate the number of corners on a particle, one has a postulate that a pyramid with one apical corner is used for simulating a corner on a particle, and the sharper the apex, the greater the reactivity of the corner. As n increases, the top of each block is so small as to be regarded as a single point. So, one can transform the vertical blocks into pyramids with one apical corner. The base angle of a pyramid, ψ, is measured as

$$\psi = \tan^{-1} \frac{A_n}{2/2n} = \tan^{-1} \frac{2A_n \cdot n}{\sqrt{2\pi}}$$

and the equation of corners becomes:

$$\Sigma n^2 g = k\Sigma n^2 \left(\tan^{-1} \frac{2A_n \cdot n}{\sqrt{2\pi}}\right)^2 = \frac{2k}{\pi^2} \Sigma n^4 A_n^2 \tag{13}$$

D. Other Orthogonal Functions

One practical implication of a complete series is that one can faithfully represent a given profile to an accuracy that is determined by the number of terms chosen to represent that profile. There are other orthogonal series but not all are complete, and this lack of completeness causes one to hesitate in testing their effectiveness in morphological analysis. An example of this type is the Radamacher function that is a square wave with a unit mark space ratio. The first five Radamacher functions are given in Figure 8. By comparison with Figure 6, Chapter 2, it will be observed that the Radaacher functions form an incomplete but true subset to the Walsh functions.

The Haar functions have the capability of representing a given function with few constituent terms to a high degree of accof the first eight Haar functions. Along with the Walsh and Fourier series, Haar functions are quite capable of representing complicated profiles as Figure 5 illustrates. In general, as one would expect, it takes less Fourier terms to represent a continuous type of wave form than the number of Walsh or Haar terms needed, respectively. Similarly, for a discontinuous wave form, the square wave sets of function are to be preferred because less of their terms will be required to synthesize such a wave form.

One might speculate that the Haar function which is a local function, could be used to represent features on the profile of a particle that occur infrequently more economically than either Walsh or Fourier (which are both global functions.)

III. THE PROBLEM OF REENTRANTS

A. The (R,Θ) Method

The (R,Θ) method is very useful for representing analytic or holomorphic particle profiles. These contain no reentrants and therefore for every value of Θ there is one,

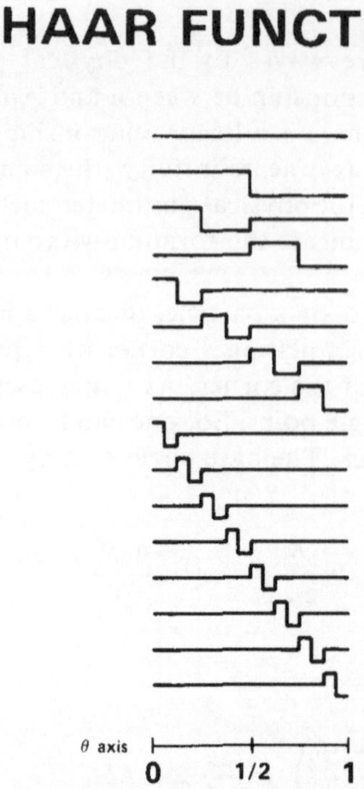

FIGURE 9. Haar functions. (From Beauchamp, K. G., *Walsh Functions and their Applications,* Academic Press, London, 1975. With permission.)

and only one, corresponding value of R. The analysis may therefore be conducted smoothly. However, if a profile contains a reentrant, then at certain values of Θ, there will be multiple values of R. An example of this type of problem is shown in Figure 10. The only way to solve this problem with the (R,Θ) method is to arbitrarily choose one R value as the one that matters and ignore the others. For example, in the case of the profile shown in Figure 10 one decision rule could be to select the largest value R as the value to compute with. This would tend to smooth out the curve artificially and so information would be lost from the analysis. The amount lost in this way would vary from particle to particle.

B. The (Φ,ℓ) Method[6]

Instead of changing (x,y) pairs into (R,Θ) pairs, the profile is parametrized by its arc length and change of slope $\Phi(\ell)$ from its starting point as in Figure 11 (i.e., (x,y) → $(\ell,\Phi(\ell))$). In order that comparisons between particles of different sizes be meaningful, the total arc lengths, or perimeters are normalized by defining

$$t = \left(\frac{\ell}{L}\right) 2\pi \tag{14}$$

$$\phi^*(t) = \left(\frac{\ell}{L}\, 2\pi\right) + t \tag{15}$$

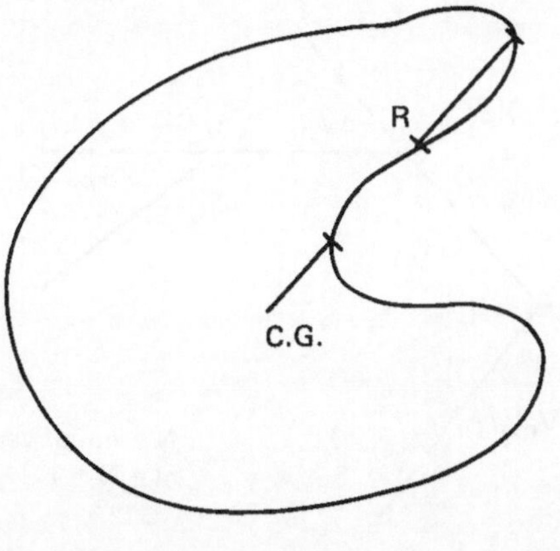

FIGURE 10. Reentrance point on particle profile. (From Fong, S. T., Beddow, J. K., and Vetter, A. F., *Powder Technol.*, submitted, 1979. With permission.)

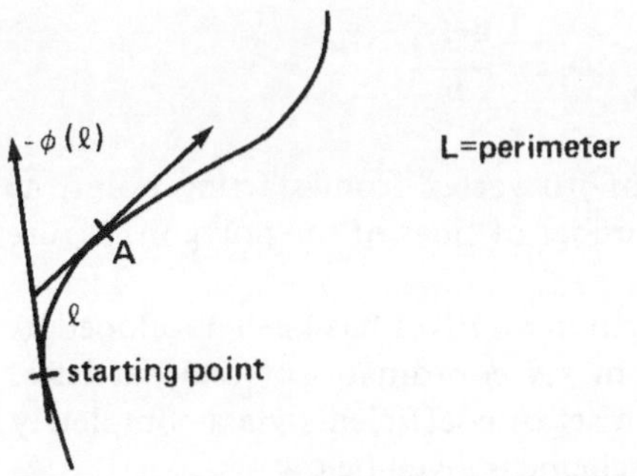

FIGURE 11. Parametric representation of a profile. (From Fong, S. T., Beddow, J. K., and Vetter, A. F., *Powder Technol.*, submitted, 1979. With permission.)

The interval of interest is $t = [0,2\pi]$ for $\ell = [O,L]$.

The function $\Phi^*(t)$ is then expanded into a Fourier series of the form

$$\phi^*(t) = \mu + \sum_{k=1}^{\infty} (A_k \cos kt + B_k \sin kt)$$

All the morphological information will be contained in the coefficients (A_k,B_k). As the ordered pair $(\ell,\phi(\ell))$ is unique for every point on the profile, multi-valueness no longer appears, and the profile is faithfully represented by (A_k,B_k).

Since all particle profiles are closed curves, they can all be approximated by polygons. If this fact is exploited, simplified equations for finding μ_o, A_k, B_k can be derived.

$$\mu_o = \pi - \frac{1}{L} \sum_{j=1}^{m} \ell_j \Delta\Phi_j \tag{16}$$

$$A_k = \frac{-1}{k\pi} \sum_{j=1}^{m} \Delta\Phi_j \sin \frac{2\pi k \ell_j}{L} \tag{17}$$

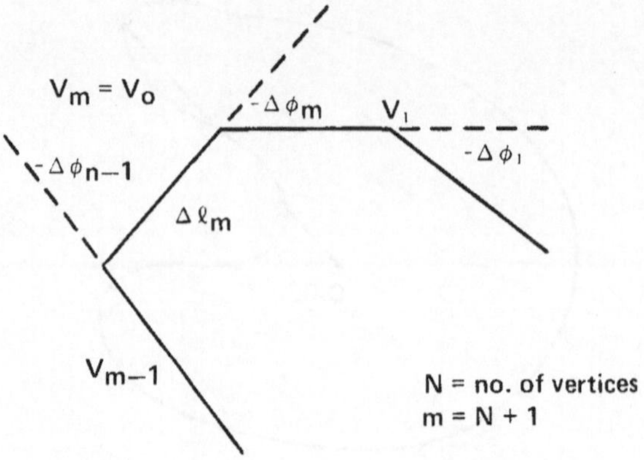

FIGURE 12. Notations of arc length and change of shape. (From Fong, S. T., Beddow, J. K., and Vetter, A. F., *Powder Technol.*, submitted, 1979. With permission.)

$$B_k = \frac{1}{k\pi} \sum_{j=1}^{m} \Delta\Phi_j \cos \frac{2\pi k \ell_j}{L} \qquad (18)$$

in which $_j$ = arc length of jth vertex from starting point, $\Delta\phi_j$ = change of slope at vertex j, and m = total number of sides of the polygon. Figure 12 illustrates the meaning of item notations.

A computer program written on PL/1 has been developed by the authors to calculate coefficients μ_o, A_k, B_k from x,y coordinates of real particles. A subprocedure which regenerates profiles from a set of coefficients via a completely different route has also been written. The overall scheme is given below.

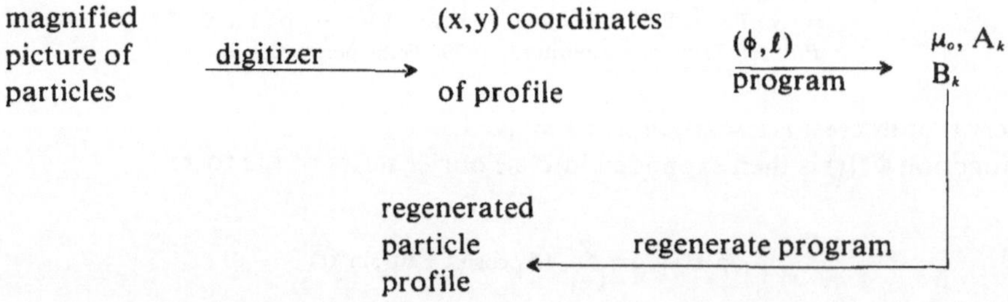

C. (RΘ) and (Φ,ℓ) compared

By comparing the regenerated profile with the original, validity and applicability of the method is checked visually. Figures 13 and 14 show some of these results and a comparison of the ϕ,ℓ method with the R,Θ method for both simple and complex particle profiles, respectively. It is evident that wherever reentrance points appear, the ($\phi\ell$) scheme is superior to the (R,Θ) one, although simple (i.e., holomorphic) profiles are adequately represented by both.

If the profile of an m-sided regular polygon is considered, it can be proved (in the next section) that A_k will vanish for any k whereas B_k is nonzero only when k is a whole number multiple of m. In addition, those coefficients that are nonzero are given by a very simple formula.

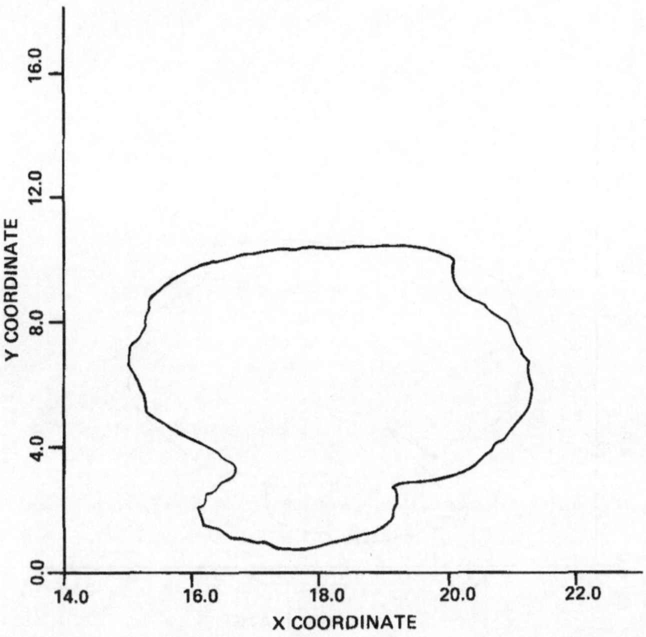

A

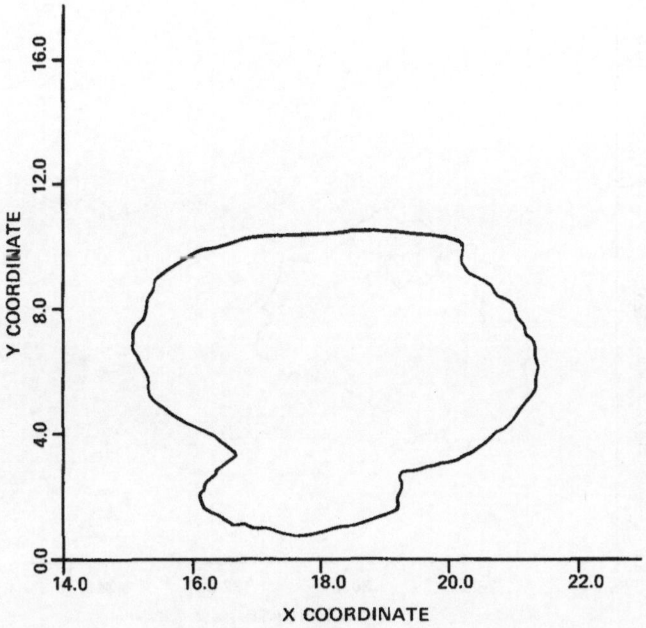

B

FIGURE 13. (A) Original shape of Cu59, (B) regenerate shape of Cu59 from (Φ, ℓ) program using 75 coefficients, (C) regenerate shape of Cu59 from (R,Θ) program using 75 coefficients. (From Fong, S. T., Beddow, J. K., and Vetter, A. F., *Powder Technol.*, submitted, 1979. With permission.)

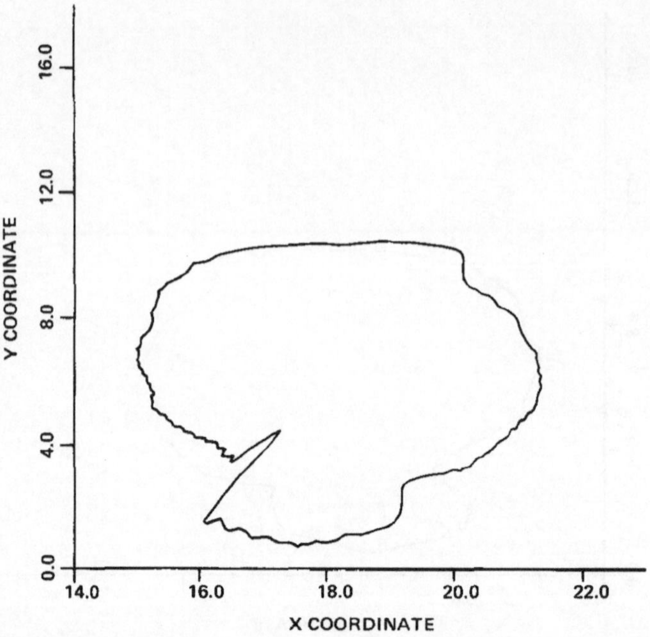

FIGURE 13C

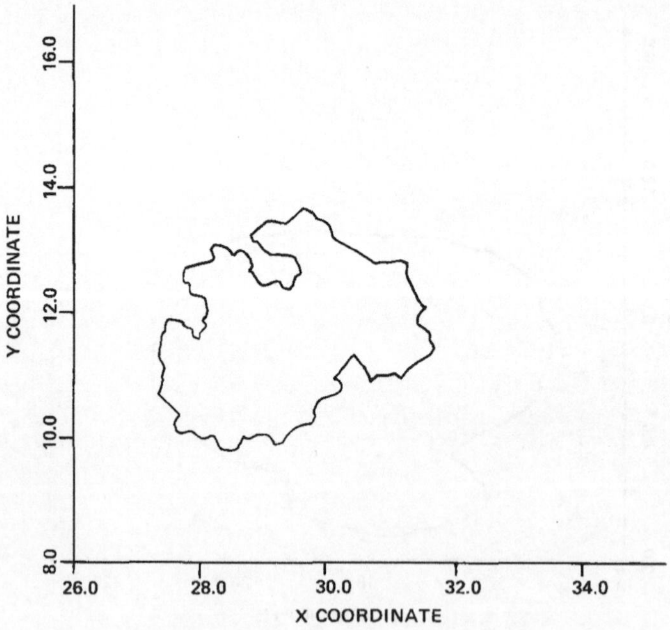

A

FIGURE 14. (A) Original shape of Cu89, (B) regenerate shape of Cu89 from (ϕ,l) program using 75 coefficients, (C) regenerate shape of Cu89 from (R,Θ) prooram using 75 coefficients. (From Fong, S. T., Beddow, J. K. and Vetter, A. F., *Powder Technol.*, submitted, 1979. With permisson.

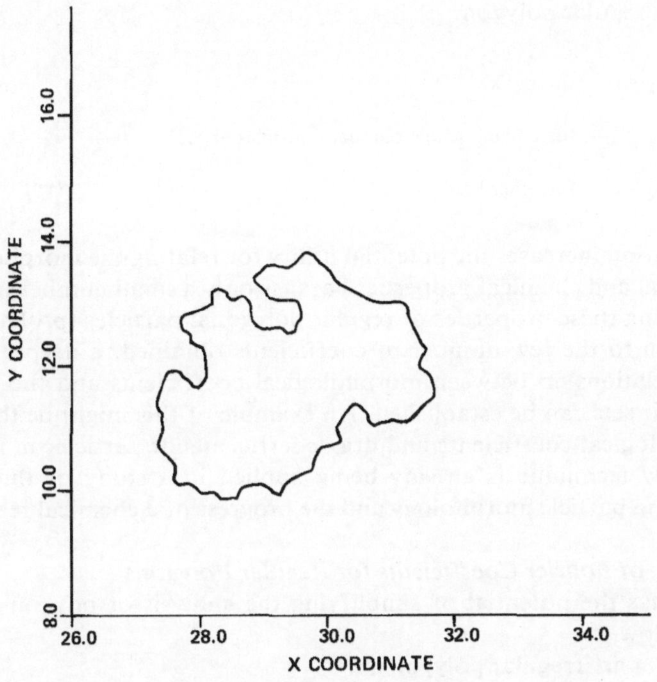

FIGURE 14B

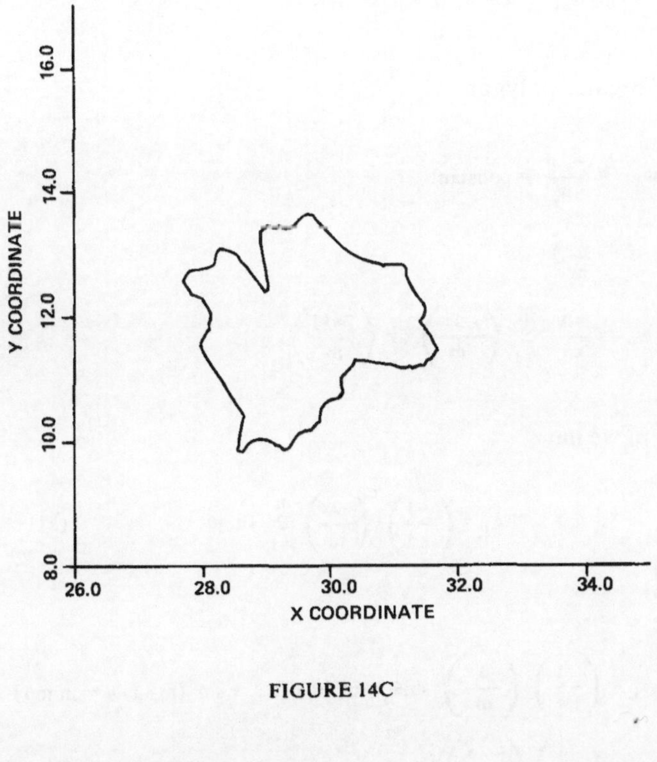

FIGURE 14C

For an m-sided regular polygon:

$$A_k = 0 \quad \text{for any k,}$$

$$B_k = \frac{-2}{k} \quad \text{for k being whole number multiples of m,}$$

$$B_k = 0 \quad \text{for other ks.} \tag{19}$$

This simplification increases the potential utility for relating the morphological coefficients to physical and chemical properties because only a small number of coefficients exist. By measuring these properties of regular polygonal particles (probably artificial) and relating them to the few number of coefficients obtained, a deeper insight as to the functional relationship between morphological coefficients and the properties of particles and their sets can be established. An example of this might be the connection between morphological coefficients and drag coefficients of particles moving through a fluid. This new technique is already being applied in a study of the relationship between changes in particle morphology and the progress of a chemical reaction.

D. Simplification of Fourier Coefficients for Regular Polygons

This method has the potential of simplifying the analysis of poygon profiles as is demonstrated below.

The equation for an irregular polygon is:

$$A_k = \frac{-1}{k\pi} \sum_{j=1}^{m} \Delta\Phi_j \sin \frac{2\pi k \ell_j}{L} \tag{20}$$

For an m-sided regular polygon:

$$\Delta\Phi_j = \frac{-2\pi}{m} = \text{constant}$$

$$\ell_j = \frac{L}{m} j$$

$$\therefore A_k = \frac{-1}{k\pi} \sum_{j=1}^{m} \left(\frac{-2\pi}{m}\right) \sin \left(\frac{2\pi kj}{m}\right)$$

Putting $\Theta = 2\pi k/m$ we have

$$A_k = \left(\frac{-1}{k\pi}\right) \left(\frac{-2\pi}{m}\right) \sum_{j=1}^{m} \sin j\theta \tag{21}$$

If m is odd:

$$A_k = \left(\frac{-1}{k\pi}\right) \left(\frac{-2\pi}{m}\right) [\sin\theta + \sin 2\theta + \ldots\ldots + \sin (m-1)\theta + \sin m\theta]$$

$$= \left(\frac{-1}{k\pi}\right) \left(\frac{-2\pi}{m}\right) [\sin\theta + \sin (m-1)\theta + \sin 2\theta + \sin (m-2)\theta + \ldots\ldots$$

$$+ \sin r\theta + \sin (m-r)\theta + \sin m\theta]$$

Using the trigonometrical identity:

$$\sin A + \sin B = 2\sin \frac{A+B}{2} \cos \frac{A-B}{2}$$

$$A_k = \frac{2}{km} \; [2\sin \frac{m\theta}{2} \; (\cos \frac{2-m}{2}\theta + \cos \frac{4-m}{2}\theta + \ldots.. + \cos \frac{2r-m}{2}\theta + \sin m\theta]$$

but,

$$\sin \frac{m\theta}{2} = \sin \left(\frac{m}{2}\right) \left(\frac{2\pi k}{m}\right)$$

$$= \sin k$$

$$= 0 \qquad \text{for any integer k}$$

Also

$$\sin m\theta = \sin m \left(\frac{2\pi k}{m}\right)$$

$$= \sin 2k\pi$$

$$= 0 \qquad \text{for any integer k}$$

Therefore every term within the bracket on the R.H.S. of the equation is zero.

$$\therefore A_k = 0 \qquad \text{for any integer k}$$

If m is even, the middle term sin m/2 is left over in Equation 1. But

$$\frac{m}{2}\theta = 0 \qquad \text{for any integer k}$$

$$\therefore Ak = 0 \qquad \text{for any integer k and m}$$

$$B_k = \frac{1}{k} \sum_{j=1}^{m} \Delta\Phi_j \cos \frac{2\pi k \ell_j}{L}$$

Substituting

$$\Delta\phi_j = \frac{-2\pi}{m}$$

$$\ell_j = \frac{L}{m} j$$

into the equation we have

$$Bk = \frac{-2}{km} [\cos\theta + \cos2\theta + \ldots$$

$$+ \cos(m-j)\theta + \cos m\theta] \ldots \qquad (22)$$

but an m-sided regular polygon has m-fold rotational symmetry. According to Equation 22, B_k will be nonzero only if k is a whole numer multiple of m or k/2 = integer, i.e., only B_m, B_{2m}, B_{3m}, etc. are nonzero. When k/m is an interger, $\theta = 2\pi k/m$ will be a whole number multiple of 2π and $\cos 2n\pi = 1$ for any n.

$$B_k = \frac{-2}{km} [1 + 1 + 1 + \ldots\ldots + 1]$$

$$= \frac{-2}{km} m$$

$$= \frac{-2}{k} \qquad (23)$$

IV. TRANSLATION INVARIANCE[7]

A. The Problem of Translation Invariance

While the discrete Fourier transform is invariant to the circular time shift of the input signal, the Walsh transform is not invariant. What this means in practice is that whereas a Fourier series can be used to represent a particle profile quite independently of the origin that one starts to digitize from, this is not the case for the discrete Walsh transform. This is shown in Table 3 and more graphically in Figure 15.[8] In Table 3 the numbers represent the discrete Walsh transform of a sinusoidal wave. There are 32 of them. In part b of Table 3 the numbers represent the same wave-form but shifted by $\pi/2$ radians. Perhaps more clearly in Figure 15 the values of the Walsh coefficients change over the interval with a 2π radian phase shift in size as well as in sign. Table 4 gives a set of Walsh coefficients for a particle profile. Note that the effect is the same for this closed form as it is for the wave-form. It is clearly unsatisfactory to use a transform that gives different results depending upon where you start to take the sample points from.

B. The R Transform

Fortunately, a form of the fast Walsh transform has been developed which is invariant to the cyclic shift of the input data. But there is a price that has to be paid for this important advantage. The original signal cannot be regenerated from the output of this transform. While this is not great disadvantage, it does compare unfavorably with the Fourier transform. However, the other part of the price that has to be paid is more serious, as shown in Figure 16,[8] the higher-order terms that the Walsh transform will pick up the detail with are simply not there if one uses the R transform. This can be a serious disadvantage if one does not take it into account. We can take note that if it was not the attractiveness of the type of modeling that the morphological characteristics represent, it would not be worth it to use Walsh or R type transforms over the Fourier type. In addition, although the Walsh transform is a little quicker to use than is the Fourier transform, as is shown in Table 5, this cannot offset the fact that finer detail of particle profiles will not be obtained. However, in the case of the Haar transform, there appears to be a substantial advantage in computation time as com-

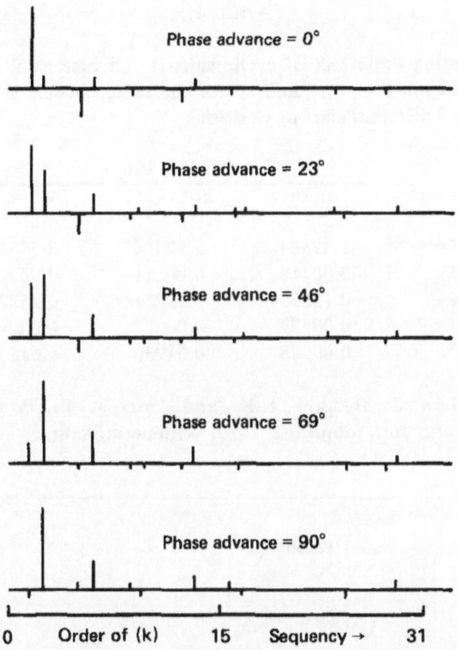

FIGURE 15. Walsh transforms of a sinusoidal wave-form with circular time shift (N = 32). (From Beauchamp, K. G., *Walsh Functions and their Applications*, Academic Press, London, 1975. With permission.)

TABLE 3

Walsh Transform Coefficient

For a Simple Sine Waveform N = 32

0	0.663	0.063	0	0	−0.263	0.025	0
0	−0.052	−0.006	0	0	−0.126	0.013	0
0	−0.013	−0.002	0	0	0.006	0	0
0	−0.025	−0.002	0	0	−0.062	0.006	0

For the Sine Waveform Shifted by 90°

0	−0.063	0.663	0	0	0.025	0.263	0
0	0.006	−0.052	0	0	0.013	0.126	0
0	0	−0.013	0	0	0	−0.006	0
0	0.002	−0.025	0	0	0.006	0.062	0

From Beauchamp, K. G., *Walsh Functions and their Applications*, Academic Press, London, 1975. With permission.

pared with the others (Table 5). The advantages and disadvantages of using the Haar transform have yet to be investigated.

TABLE 4

Illustrating the effect of cyclic shift in the case of Walsh Coefficients 1, 3, 4, 7 and 10 for the same particle in the case of 3 different starting positions

	Starting Point		
Coefficient	Position 1	Position 2	Position 3
1	2.57564	2.57564	2.57564
3	0.00418	0.00111	0.00577
4	0.18598	0.19251	0.15227
7	0.00529	0.00722	0.00049
10	0.00988	0.01936	0.01330

From Lew, G., Beddow, J. K., and Vetter, A. F., *Powder Metallurgy Int.*, submitted, 1979. With permission.

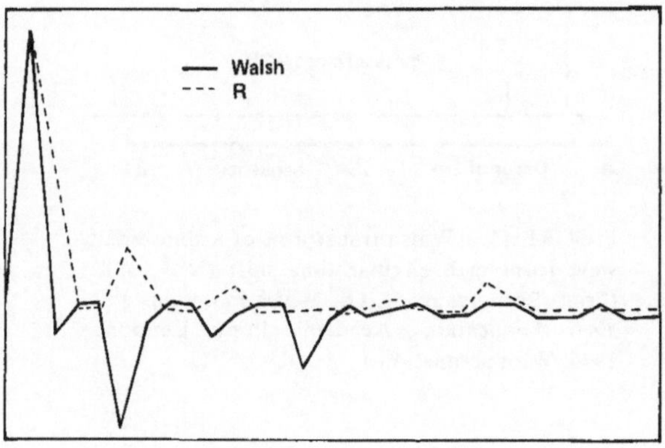

FIGURE 16. Comparative examples of the R-transform and the Walsh transform. (From Beauchamp, K. G., *Walsh Functions and their Applications*, Academic Press, London, 1975. With permission.)

TABLE 5

Fourier, R, and Haar compared

Transform	Time(s)	Data storage (N = 1024)
Fourier	9.48	4K
Walsh	1.60	3K
Walsh—Ulman(R)	1.25	4K
Haar	0.29	4K

From Beauchamp, K. G., *Walsh Functions and their Applications*, Academic Press, London, 1975. With permission.

V. METHODS OF FEATURE EXTRACTION

At its present stage of development, morphological analysis of fine particles transforms a set of x,y coordinates into a set of coefficients (Fourier, Walsh, Haar, etc.). These coefficients may then be further analyzed to yield more condensed forms of information. These latter forms may in turn be input for further analysis as required. For convenience, we have divided the discussion of these methods of analysis into two sections. This one and the section following.

A. Harmonic Analysis

Ehrlich and co-workers have developed a sophisticated set of techniques for the harmonic analysis of sediments.[9,10] After digitizing the particle profile and subjecting the (x,y) data to Fourier analysis, the sets of coefficients may be displayed as one of the following:

1. Mean harmonic amplitude spectra,
2. Shape frequency distributions.

An example of a mean harmonic amplitude spectrum is given in Figure 17 for three different sediments: Brown Mountain Granite, Piedmont Granite, and Felsic Tuff.[10] As can be observed, there are differences between these three sediments over a broad range of harmonics. In cases where these differences in harmonic amplitude spectra are insignificant, nonexisting, or just confusing, it is possible to develop shape frequency distributions with a view to clearly distinguishing between morphologies. Figure 18 shows an example of this technique for the three sediments listed above using the 14th harmonic.[10] It will be seen that there are few high values of the 14th harmonic in the sample of Brown Mountain Granite. There are, however, high values of the 14th harmonic in the sample of the Piedmond Granite.

It is possible to develop algorithms in order to distinguish between samples of sediments with different morphologies.[10] For example, in Figure 19, there is a schematic illustration of an algorithm for distinguishing the sediments named. As one can see, different legs of the investigation use quite different harmonics in order to classify the data.

Ehrlich and his co-workers[9] have developed a very interesting set of techniques for harmonic analysis of fine particle profiles. This set consists of three techniques (these are the main parts): The X^2 test, R-mode factor analysis, Q- mode factor analysis.

In essence the X^2 test yields information concerning the most important coefficients with respect to discriminating between fine particle sets. The X^2 test therefore serves as a spot test. The R-mode factor analysis generates basic factors that are linearly related to the significant coefficients determined from the X^2 test. The R-mode analysis is therefore equivalent to a qualitative analysis. The Q-mode factor analysis yields the proportions of particle shapes in a mixture of shapes. This is a quantitative analysis technique. This powerful set of techniques has been used to demonstrate the origin of sand in river estuaries and the origin of littoral sands off the coast of the U.S.

Lewis and Goldman have used the Kolmogorov statistic for comparing distributions.[25] The use of this statistic in place of the X^2 will be investigated.

B. Signature Analysis

This idea has been proposed and largely developed by Meloy.[5a] If the set of Fourier coefficients corresponding to a holomorphic particle profile is plotted either as: a graph of A_n versus log n, or of log A_n versus log n, or of log A_n versus log 1/n (that is

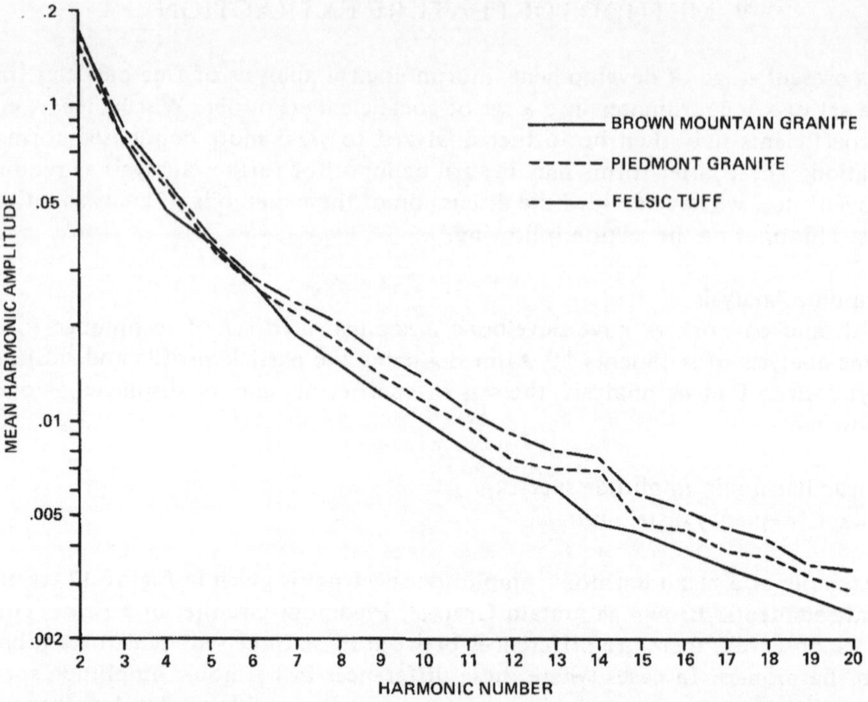

FIGURE 17. Mean harmonic amplitude spectra for Brown Mountain granite, Piedmont granite, and Felsic tuff. (From Pryzgocle, R. S., Master's thesis, University of South Carolina, Columbia, 1976. With permission.)

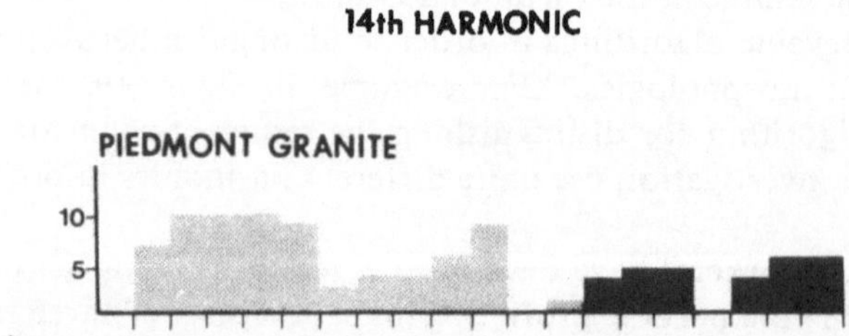

FIGURE 18. Discrimination of the Piedmont from the Brown Mountain Granite based upon shape-frequency histograms at the 14th harmonic. (From Pryzgocle, R. S., Master's thesis, University of South Carolina, Columbia, 1976. With permission).

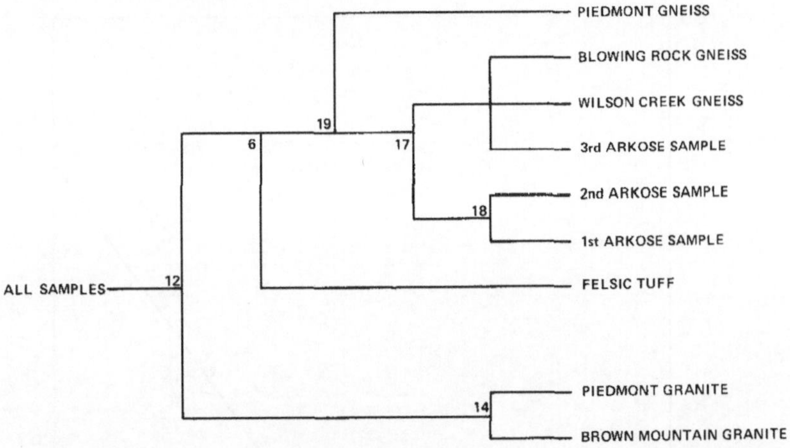

FIGURE 19. Rock discrimination algorithm, contrasting all nine samples collected over single lithologies using harmonics 14, 12, 6, 19, 17, 18. (From Pryzgock, R. S., Master's thesis, University of South Carolina, Columbia, 1976. With permission).

log λ). The resultant curve decreases as n increases. One example of a signature is given in Figure 20 for an atomized copper-lead particle. Another idea behind this concept is that particles with similar signatures should lie within the same envelope. This is illustrated in Figure 21. One can hypothesize, for example, that particles of the same substance with different signature envelopes may have originated by different mechanisms.

In the case of a Fourier representation of a particle profile, three different types of particle signature have been proposed each corresponding to a certain type of random particle. These are: type I — in which only the ∝s vary randomly (∝ is the phase angle), type II — in which both the Ans and ∝s vary randomly, type III — in which the Ans vary randomly. The particle signature was hypothesized as a straight line such that its equation is [5]:

$$A_n = A_I(n)^s \tag{24}$$

in which A_I is the intercept of the ln(An) axis; s is the slope of the line.

1. Variation of αn (Type I Randomness)

A pseudorandom number generator was used to vary the values of αn and An randomly. The varieties generated by RANDNO appear to be uniformly distributed on the interval (0,1). When called each time, this generation function returns a distinct six-digit number between 0 and 1. If multiplied by 360, a value returned by RANDNO can be used to convert the value of αn by the equation:

$$\alpha'n = n + 360 \times \text{RANDNO} \tag{25}$$

which means that α′n varies from 0° to 360° randomly. This uniformly distributed random α′n, along with the unchanged An, was used to regenerate the particle shapes by the definition of type I random particle.

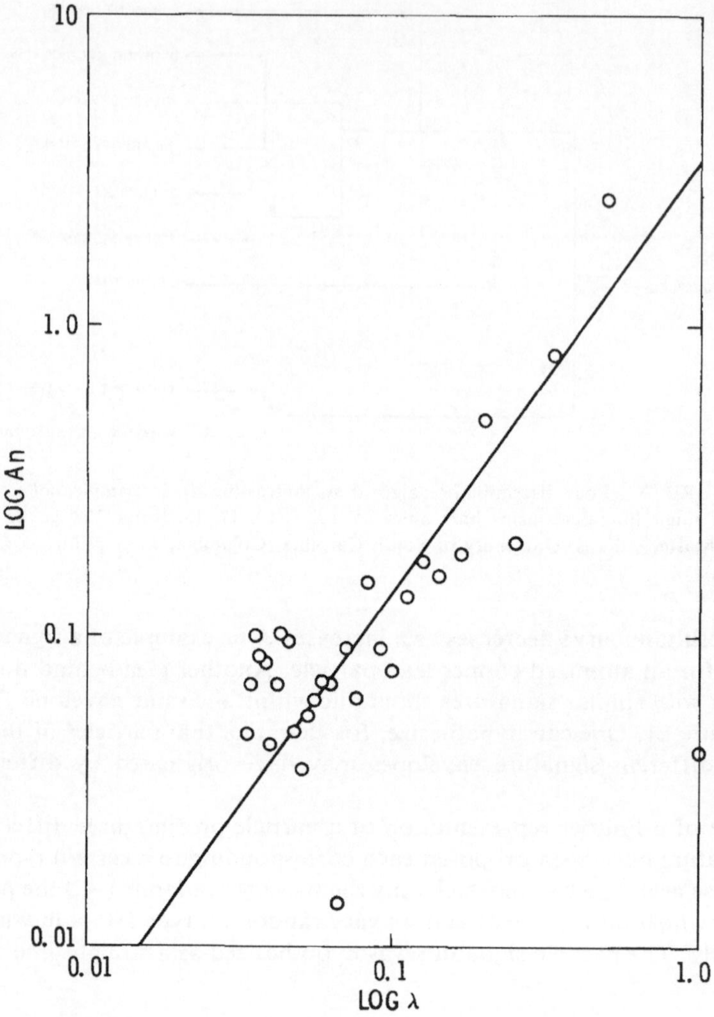

FIGURE 20. Particle signature. Log An vs. Log λ, Cu-Pb, Slope = 1.39.

A set of five original profiles of atomized copper-lead particles is shown in Figure 22 and the artificial regenerates of the type I random particles are shown in Figure 23 with a commentary given in the left-hand side of Table 6.

2. Variation of An and αn Simultaneously

In type II, the RANDNO function was supplied to produce the randomness of An in addition to αn. In this case, An was assumed to have a variation of 50% of itself. Therefore, An varies within 1.0 to 1.5 of its original size as following:

$$An = An + 0.5 \times RANDNO \times An \tag{26}$$

A set of type II random regenerates of the five Cu-Pb particles is given in Figure 24 with the comment in the right-hand side of Table 6.

Although these examples are interesting in that they show visually some effects due

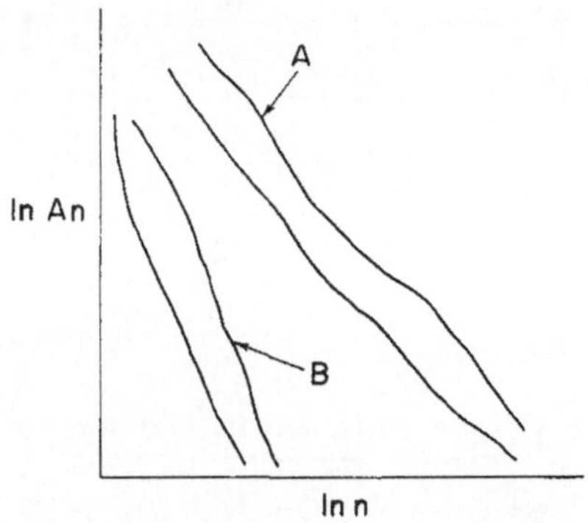

FIGURE 21. Signatures of two particle sets (A and B). (From Lew, G., Beddow, J. K., and Vetter, A. F., *Powder Metallurgy Int.*, submitted, 1979. With permission.)

to different types of randomness, it is clear that a much more basic approach will be needed in order to develop a useful signature theory. Fortunately, this line of inquiry is currently being pursued.[11]

C. Feature-Space Representation of Particle Shape[12]

In order to be able to represent the complex patterns of relationships to be observed in particle morphology analysis, it is necessary to develop new ways of representing these relationships. A recent paper discusses a prototype system of feature-space representation.[12] The new method is a three-dimensional plot of form and roundness. A triaxial graph in the plane of the paper is used by the observer to plot the form characteristics of the particles under study. These three characteristics are thickness, breadth, and length, respectively. The axial lengths are scaled from zero to one such that the following relationship holds:

$$(L + B + T = 1) \tag{27}$$

This method of graphical representation has been used to analyze form characteristics of samples of sand and SiC. The results are sets of contours as shown in Figure 25.[12] The roundness of the particles is represented in the vertical direction as shown in Figure 26.[12] Those familiar with ternary phase diagrams will recognize the inherent difficulties which one meets when using three-dimensional plots to represent complex systems. However, one must remember that complicated and cumbersome though the 3-D plot may be, it constitutes a distinctly simpler generalization than the plethora of data that it represents. This new method is therefore a significant advance for particle morphology analysis; it points a way that representational methods might go, and it also highlights the need for sophisticated computer interfaced imaging systems in order that more data can be more easily amassed and understood.

VI. DETERMINISTIC, STATISTICAL, AND FUZZY CLASSIFIERS

In order to pursue the desired objective of relating the properties of particulate matter to the morphological characteristics of the particles therein, we have to try to un-

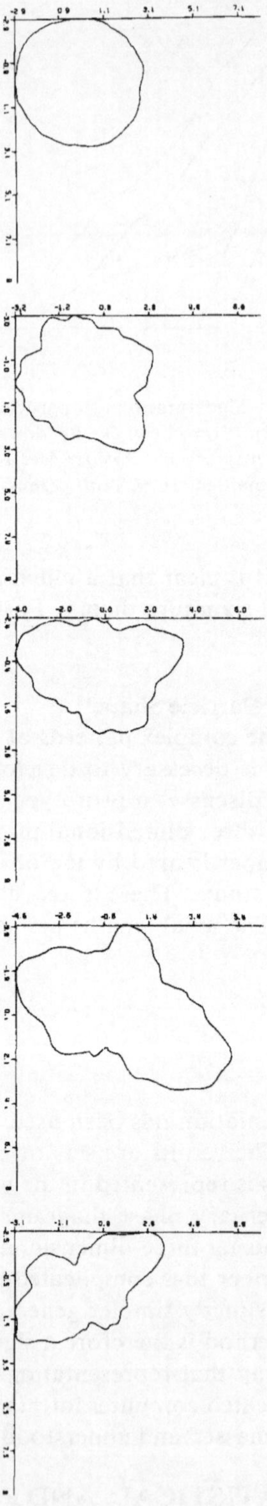

FIGURE 22. Five CuPb particle profiles.

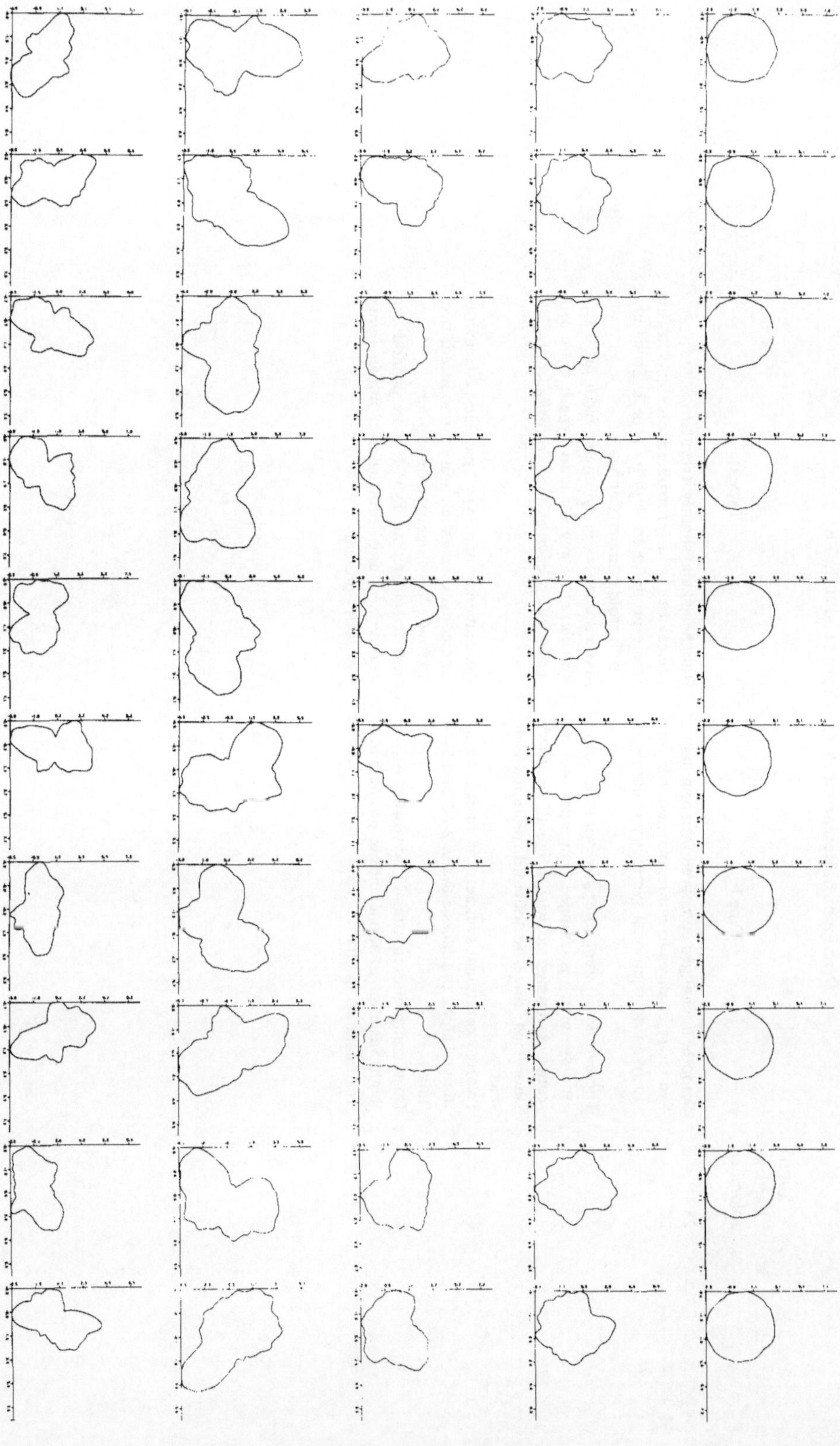

FIGURE 23. Type I regenerates.

TABLE 6

Description of Regenerate Profiles Type I and II Cu-Pb

Cu-Pb Particle (Number)	Type I	Type II
28	Smooth, unchanged compared with original	Roughened and slightly irregular
34	Symmetry changes in nonregular manner four to five-fold rotational symmetry is dominent	Concavities and protruberences tend to be more pronounced than in Type I. Still four- to fivefold symmetry apparent
33	These look different compared with each other and symmetry is changed; more pronounced concavity compared with original; long side shortened; three- to fourfold rotational symmetry	Marked changes in symmetry from profile to profile; more angular than Type I; more pronounced concavities and protrusions also
47	Twofold rotational symmetry is strong; variation of alpha produces either a 2—3 or a 2—4 mix	More angular than Type I particles; concavities and convexities much more pronounced; two- to fourfold symmetry more definite
31	Three concavities in original; low order An are high; variation of alpha tends to promote mirror-image symmetry	More irregular than Type I; more pronounced concavities and protrusions; mirror-image symmetry present

FIGURE 24. Type II regenerates.

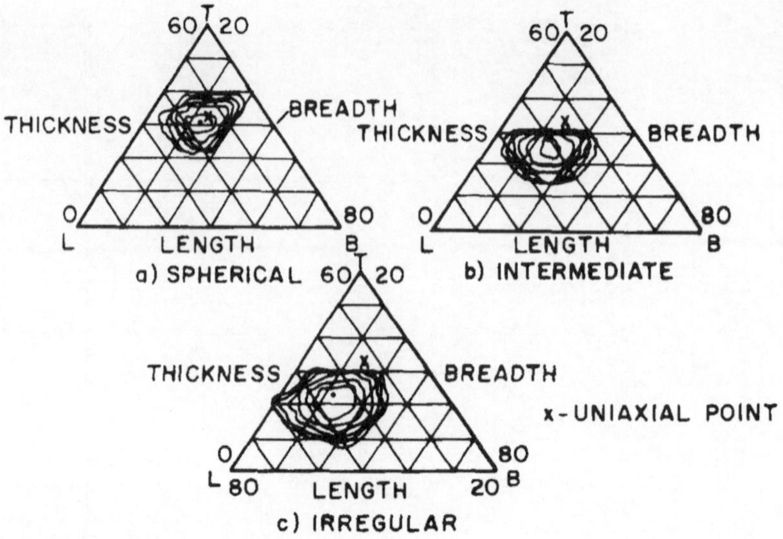

FIGURE 25. Triaxial contour diagrams of shape-sorted sands. (From Davies, R., *Powder Technol.*, 12(12), 111, 1975. With permission.)

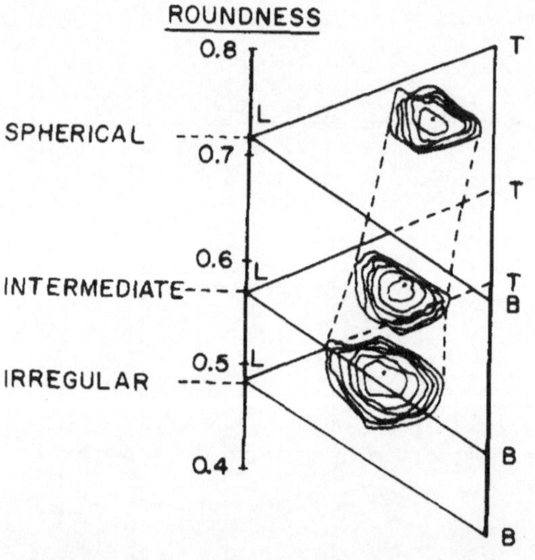

FIGURE 26. Form-roundness feature-space representation of shape-sorted sands. (From Davies, R., *Powder Technol.*, 12(12), 111, 1975. With permission.)

derstand many of the quite sophisticated methods of pattern recognition which have been developed during the last two decades. When we understand them, we have to adapt them to our own use. Some of the work reported here is part of that ongoing process. A useful starting point for pattern recognition is given in Reference 13. Reference 14 gives a list of articles on fuzzy sets.

The ultimate test of this approach lies in the ability to predict both particle behavior and macroscale properties of particle sets in engineering applications and to predict and control morphic properties of particle sets as they are produced. It is hoped and expected that such an ability will follow the understanding of morphological analysis.

Pattern recognition techniques have been shown to have potential utility in particle shape analysis.[15] In particular five methods have been identified in this earlier study. These methods are

1. Classification by distance functions
2. Classification by likelihood functions
3. Classification by trainable classifiers — deterministic methods
4. Classification by trainable classifiers — statistical methods
5. Classification by syntactic methods

The concept of classification by distance functions involves the use of a distance criterion. The item to be classified is incorporated in the pattern space along with the previously established training pattern classes. The class that is closest to the item to be classified is judged to be the class to which the item belongs. The use of likelihood functions requires the knowledge of the probability of occurrence of a particular pattern class and of the loss or gain associated with the classification of each element. A deterministic pattern classifier makes no assumptions regarding the statistical properties of the pattern class. Instead, the decision functions are generated by means of iterative algorithms. In the case of a statistical pattern classifier, the purpose of the method is to minimize the probability of error (i.e., misclassification). Syntactic methods of pattern recognition seek to develop a set of rules with which to generate the pattern from a given starting set of pattern primitives. Unlike the other methods, this one gives a physical picture of the pattern.

It appears to be appropriate to use a deterministic classifier in the initial design/training stage of the development of an analytical capability for morphological analysis. This has been done in the following:

1. In the case of four metal powders using a form of Equation 11 in which α is zero and β is unity. The summation was arbitrarily split into three groups.
2. In the case of four metal powders with stereological-like descriptors using a Quantimet 720 System,[16]
3. The deterministic classifier was used in an experiment to classify the same four particle sets using Walsh coefficients as pattern features and in another experiment using morphological characteristics as features.[17]

The use of statistical classifiers has been studied in which the features used were Walsh coefficients. Three classification rules were utilized:[18] the Linear Discriminant Function, the Quadratic Discriminant Function, and the Fix-Hodges Rule.

What has not yet been studied and reported is the use of fuzzy classifiers, although some of the properties and potential applications of fuzzy sets have been reported.[19] Two interesting models have been proposed as follows:

The first model leads to a nonfuzzy[20] or a fuzzy[21] ISODATA algorithm depending on the value of m.

$$J_m(U,\underline{v}) = \sum_{k=1}^{n} \sum_{i=1}^{c} (U_{ik})^m \, || \underline{X}_k - \underline{V}_i ||^2 \quad 1 \leq m < \infty \qquad (28)$$

where J_m is the generalized total squared error incurred by representing each fuzzy subset by a fuzzy cluster center $\underline{v}$. One seeks to minimize this error. c is the number of fuzzy subsets, n is the number of features, U is the fuzzy subset, x is the feature vector. If m is unity, then J_1 is the well-known within-group sum of squares. The case where m equals 2 has been investigated using the classic data of Anderson[22] which is a set of 150 four-dimensional vectors in one of three subspecies of Iris.[23] The analysis was highly satisfactory and this indicates a similar prognosis in morphological analysis of particulates.

The second model suggests ways in which the problem of formulating a total representation of physical properties of a particulate might be approached. Thus[24]

$$f_A(x) = \sum_{i=1}^{n} c_i^* \phi_i(x) \tag{29}$$

where f_A is the membership function of a fuzzy set A in Σ^n, c_is are undetermined constants, $\phi_i(x)$ are known linearly independent, bounded functions of the vector x in Σ^n.

A. A Simple Decision Function for Two Pattern Classes

This experiment was conducted on a small number of particles of two types and two sieve sizes: atomized Cu-Pb and sponge Fe particles, a coarse fraction (+170 mesh) and a fine fraction (−400 mesh). The purpose of the experiment is to illustrate the potential utility of even a simple classifier.

B. Features

The following features were utilized: the centroid aspect ratio (CAR) which is defined as:

$$\mathrm{CAR} = \frac{2A_0 + 2\sum_{n=1}^{n/2} a_{2n}}{2A_0 + 2\sum_{n=1}^{n/2} (-1)^n a_{2n}} \tag{30}$$

where n = the number of coefficients.

This is the ratio of the maximum dimension through the centroid divided by the dimension perpendicular to same. It is termed the centroid aspect ratio, CAR.

Three other shape characteristics are measured on a scale of gradually increasing resolving power. The set of Fourier coefficients is divided arbitrarily into three portions:

1. The sum of the first n_1 members will be characterized by the term lumpiness.
2. The sum of the intermediate members n_2 will be characterized by the term roughness.

The following attempts to develop the general formula for the summation of coefficients. The summaries are, respectively:

$$L_k = \sum_{n=1}^{n=n_1} A_n \tag{31}$$

$$R_k = \sum_{n=n_1}^{n=n_2} A_n \qquad (32)$$

In this analysis, CAR was determined using the first 30 coefficients; L was determined using the first 8 coefficients; R was determined using coefficients 9 through 15; L and R excluding A_0. Results are given graphically in Figures 27[4] and 28[4] for the pairs of features named.

Table 7 shows the number of two-dimensional items misclassified for a small number of trial-and-error linear-decision functions. The table should be read in conjunction with the appropriate graph, as designated in the table.

The choice of decision function in any one case is illustrated in the case of lines 1 and 2 of Figure 27. If one chooses line 1, then six Cu-Pb items are misclassified and only one Fe item is misclassified. If line 2 is chosen instead of line 1, then there are only two Cu-Pb items misclassified but there are now two Fe items misclassified. The investigator has to make the judgment as to which is the most advantageous choice. The choice in this example may be stated — minimize the total number of items misclassified, with either an added restriction on the Fe misclassification, or no restriction on the Fe misclassification.

These results demonstrate that it should be possible to abstract features from mathematical descriptors of particle profiles which in themselves may have physical significance in terms of the human perception of shape.

C. Deterministic Trainable Pattern Classifier[17]

The deterministic trainable pattern classifier is a set of linear mathematical equations. Each equation is a decision function. The objective of the classification process is to yield decisions concerning the pattern class* into which the specific pattern** can be classified. There are two main steps:

1. Collect measurements from the patterns to be recognized and arrange these measurements in pattern vectors
2. Classify (i.e., group into appropriate sets) the pattern vectors

The objectives of this study were:

1. To learn more about the capabilities of pattern classifiers with respect to morphological analysis of fine particles.
2. To evaluate the feasability of using morphological characteristics and also of using Walsh coefficients and Fourier coefficients as features.***
3. To compare the morphological characteristics with stereological-like properties of the same particle set.
4. To evaluate and minimize likely sources of error and information loss or distortion associated with our procedures.

The stereological-type measurements were obtained from images of particles projected onto the screen of a Quantimet® 720 Image Analyzing System.

* Pattern class: a category determined by some common attributes.

** Pattern: the description of any member of category representing a pattern class.

*** Features: individual parameters that make up the pattern.

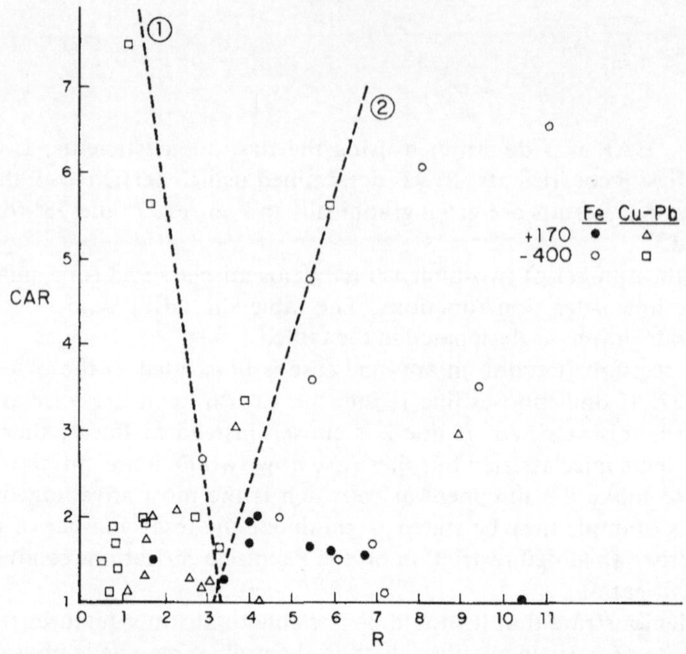

FIGURE 27. Car vs. R. (From Beddow, J. K., Philip, G. C., Nasta, M. D., and Vetter, A. F., *Powder Technol.*, 18, 19, 1977. With permission.)

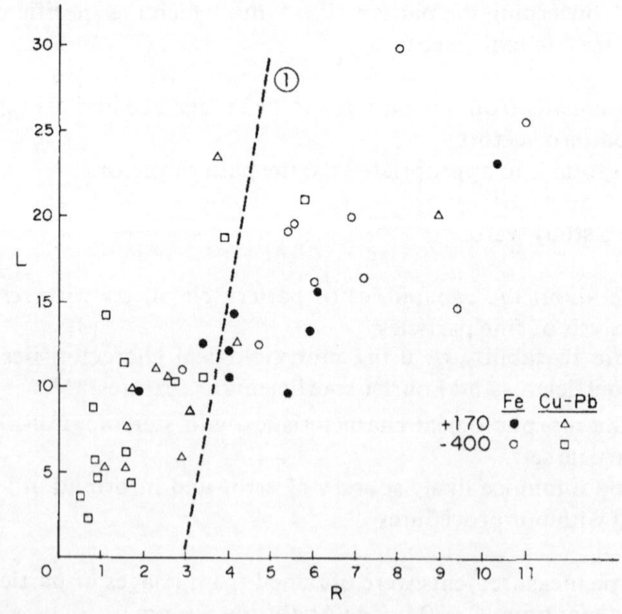

FIGURE 28. L vs. R. (From Beddow, J. K., Philip, G. C., Nasta, M. D., and Vetter, A. F., *Powder Technol.*, 18, 19, 1977. With permission.)

TABLE 7

Trials to find Linear Decision Functions from Figures 27 and 28

Topic	Items Misclassified			
	Line 1		Line 2	
	Cu-Pb	Fe	Cu-Pb	Fe
CAR vs. R	6	1	2	2
L vs. R	3	3		

From Beddow, J. K., Philip, G. C., and Nasta, M. D., *Planseeber. Pulvermetall.*, 1976.

The classification of the pattern vectors was carried out using a deterministic trainable pattern classifier. This classifier is a mathematical analogue which does not use the statistical properties of the pattern class. The word trainable in the title means that the classifier is established using known pattern classes, by means of a learning process that takes place only during the design phase of the pattern recognition system. Once acceptable results have been obtained with the training set of patterns, the system is applied to the task of actually performing recognition on samples drawn from the environment in which it is expected to operate.

Samples of four powders (Figure 29) were selected for this study: gas atomized Cu-Pb, sponge Fe, electrolytic Fe, and electrolytic Cu. The study is based upon the following two hypotheses:

1. Profiles of a given material produced by different methods possess different shapes.
2. Particles of different materials produced by a given method possess different shapes.

1. Feature Selection

Any pattern which can be recognized and classified possesses a number of discriminatory properties or features. The first step in any recognition process is to consider the problem of what discriminatory features to select and how to measure these features. The number of features needed to successfully perform a given recognition task depends on the discriminatory qualities of the chosen features.

Feature selection and extraction play a central role in pattern recognition. It is extremely difficult, if not impossible, to find the complete set of discriminating features for a pattern class.[13]

In particle characterization research, two types of features are being looked into, morphological and stereological. The former maps the outline of the projected image of a particle into an equation, while the latter characterizes a particle by its geometrical structures. Experimentally it is very difficult to take three-dimensional measurements of a particle, so its two-dimensional projected image is generally used.

2. Stereological-Type Features

Projections of a hundred particles of each type of powder were analyzed by a Quantimet® 720. The following measurements (Figure 30)[17] were obtained:

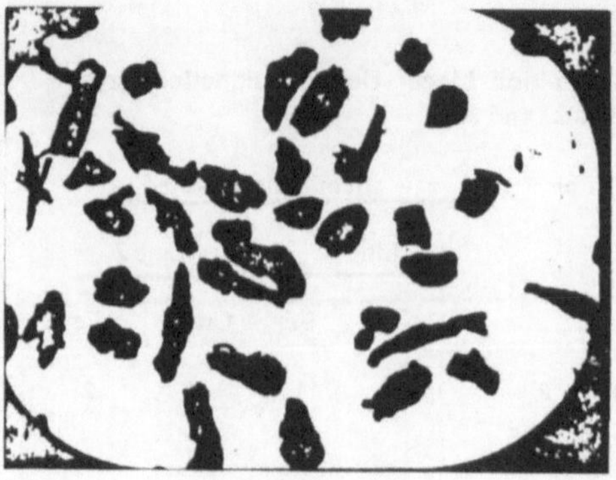

A

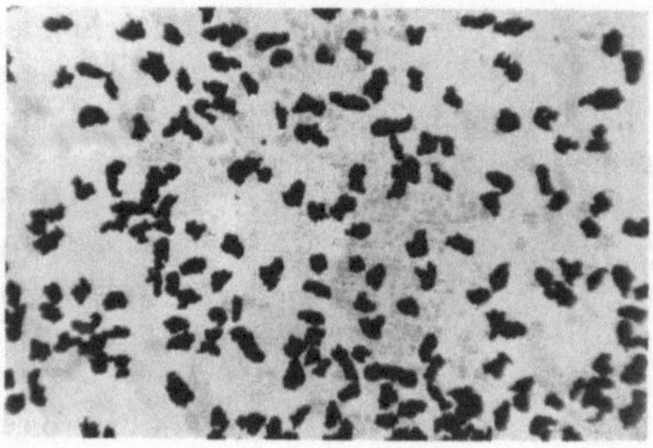

B

C

FIGURE 29. Photomicrographs of particle profiles. (A) electrolytic Fe + 170, (B) sponge Fe + 170, (C) electrolytic Cu + 170, (D) Cu-Pb + 170. (Magnification × 60). (From Beddow, J. K., *Powtech 1977, Proceedings of the 4th International Powder Technology and Bulk Solids Conference*, Heyden & Sons, London, 1977, 99. With permission.)

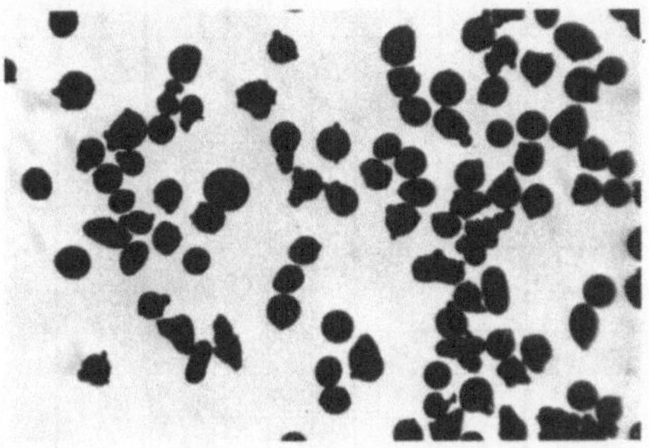

FIGURE 29 D

1. Area, A
2. Perimeter, L
3. Number of verticle intercept, I_v
4. Horizontal feret diameter, F_L
5. Number of horizontal intercept, I_h
6. Verticle feret diameter F_v

From these, five dimensionless features that are variants of shape and particle size, were derived. They are either average values of a group of particles or values derived from the use of quantitative stereological methods. These features are

$$\begin{aligned} AR &: \bar{F}_v/\bar{F}_h \\ SF1 &: \bar{A}/\bar{L}_3^2 \\ SF2 &: \bar{A}/\bar{L}^2 \\ SF3 &: \bar{I}_v/\bar{F}_h \\ SF4 &: \bar{I}_h/\bar{F}_v \end{aligned} \tag{33}$$

Two size-dependent features: $\bar{r}$, mean radius; $\bar{L}_3$ mean intercept length were also included.

$$r = \sqrt{\bar{A}/\pi}$$

$$L_3 = \tfrac{1}{2}\ \frac{\bar{A}}{\bar{I}_h} + \frac{\bar{A}}{\bar{I}_v} \tag{34}$$

3. Morphological Characteristics

In this analysis the particle profiles are modeled using square-waved Walsh functions. The properties of five morphological characteristics are investigated. These are: arc length, excess surface area, edge length, volume of the positive blocks, activity corners. These are defined in II B1 and II C.

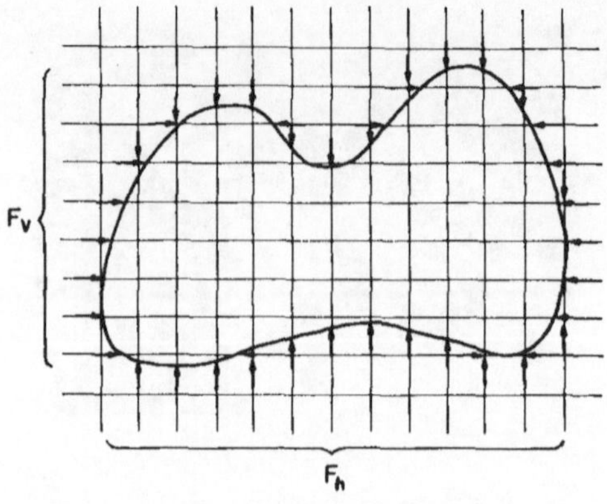

FIGURE 30. Illustrating four measurements of a particle profile. (From Beddow, J. K., *Powtech, 1977, Proceedings of the 4th International Powder Technology and Bulk Solids Conference,* Heyden & Sons, London, 1977, 99. With permission.)

The digitized particle profiles were transformed into a set of polar coordinates (R,Θ) with the centroid as origin. The function R(Θ) is then described in terms of a Walsh series as

$$R(\theta) = A_0 + A_1 \,\mathrm{WAL}\,(1,\theta) + A_2 \,\mathrm{WAL}\,(2,\theta) + \ldots \qquad (35)$$

where As are the Walsh coefficients and WAL(n, Θ) n = 1,2,3,...are Walsh functions.

This transformation can be done by using a discrete Walsh transform, but this is variant with respect to the phase of the input signal. Consequently, a circular shift of starting points brings about a different set of Walsh coefficients.

Since we wish to use Walsh coefficients as features in pattern recognition and morphological characteristic functions, the transform has to be independent of the phase or circular shift of the input signal in order to give meaningful results. A modified form of Walsh transform, known as R transform may be used. The modification involved the replacement ofthe subtractive terms obtained during the interim calculations by their absolute values. This transform is invariant to the cyclic shift of the input data. Its disadvantage is that rather less information is available concerning the sequency* distribution of the transform coefficients. In particular a lack of detail is apparent in the upper sequency range. That is to say, the transform fails to accurately represent the finest details of the particle profile. Unlike the Walsh transform, the R transform does not permit the original series to be recovered by a second transform, i.e., it is not its own inverse. However, these penalties do not limit its usefulness for our particular purpose.

4. Decision Function

In Figure 31 two hypothetical pattern classes are shown. They are conveniently separated by a line.

Let

* Sequency: is used to describe a periodic repetition rate which is independent of wave-form. It is defined as one half of the average number of zero crossings per unit time interval.

$$d(\vec{x}) = w_1 x_1 + w_2 x_2 + w_3 = 0 \tag{36}$$

be the equation of the separating line where the ws are parameters and x_1, x_2 are the general coordinate variables. From the figure a pattern x that belongs to a class ψ_1 will yield a positive quantity when substituted into d(x). Similarly, d(x) becomes negative upon substitution of any pattern from ψ_2. Therefore, $d(\vec{x})$ can be used as a decision function since, given a pattern x of unknown classification, x can be said to belong to ψ_1 if $d(\vec{x}) > 0$, or to ψ_2 if $d(\vec{x}) < 0$. If the pattern lies on the separating boundary (i.e., the line defined by Equation 36, the indeterminate condition $d(\vec{x}) = 0$ is obtained.

These concepts need not be restricted to two classes. Once a specific decision function (or functions of more than two classes are involved) has been selected, the problem becomes the determination of the coefficients in the equation.

For the n-dimensional case, the general linear decision function is of the form:

$$d(\vec{x}) = w_1 x_1 + w_2 x_2 + \ldots.. + w_n x_n + w_{n+1} \tag{37}$$

which can be expressed in vector form:

$$d(\vec{x}) = \vec{w}^t \vec{x} \tag{38}$$

where $\vec{x} = (x_1, x_2, \ldots, x_n, 1)'$ and $\vec{w} = (w_1, w_2, \ldots, w_n, w_{n+1})$ are called the augmented pattern and weight vector, respectively. And t indicates the transpose.

In this study, each pattern class is considered apart from all other classes by a simple decision surface. There an M decision functions with the property

$$d_i(\vec{x}) = \vec{w}_i^t \vec{x} = \begin{cases} >0 \text{ if } x \epsilon \psi_i \\ <0 \text{ otherwise} \end{cases}, i=1,2,\ldots M \tag{39}$$

where $\vec{w}_i = (w_{ii}, w_{i2}, \ldots, w_{in}, w_{in+1})'$ is the weight vector associated with the ith decision function.

To determine the decision function for the ith pattern class, the scheme is considered as a two-class problem ψ_i and ψ_i, where ψ denotes all classes except ψ_i. If the patterns of ψ_i are multiplied by -1, the equivalent condition $w'x > 0$ for all patterns is obtained.

The problem reduces to that of finding a vector w such that the system of inequalities $Xw > 0$ is satisfied,

$$\vec{X} = \begin{matrix} \vec{x}_1^t \\ \vec{x}_2^t \\ \vdots \\ \vec{x}_n^t \end{matrix} \tag{40}$$

The learning algorithm used here to determine $\vec{w}$ belongs to the gradient technique.

Basically, gradient schemes provide a tool for finding the minimum of a function. Upon it, many iteratve schemes have been devised.

If a certain criterion function* is chosen so that it achieves its minimum value whenever $\vec{w}'\vec{x}_i > 0$, then finding the minimum of the function for all i, i = 1,2,...,n is equivalent or solving the given system of linear inequalities. Different algorithms use different criterion functions.

The general gradient descent algorithm may be written as

$$\vec{w}(k+1) = \vec{w}(k) - c \left. \frac{\partial J(\vec{w},\vec{x})}{\partial \vec{w}} \right|_{\vec{w}=\vec{w}(k)} \quad (41)$$

where $\vec{w}(k+1)$ is the new value of $\vec{w}$, $c > 0$ distates the magnitude of the correction, $J(\vec{w},\vec{x})$ is the criterion function which is chosen with the situation. $\frac{-\partial J}{\partial \vec{w}}$ is the negative of the gradient points in the direction of the maximum rate of the decrease of J.

$$J(\vec{w},\vec{x},\vec{b}) = \frac{1}{2} \sum_{j=1}^{N} (\vec{w}^t \vec{x}_j - b_j)^2 = \frac{1}{2} \| X\vec{w} - b \|^2 \quad (42)$$

where $\vec{b} = (b_1, b_2, ..., b_N)$ are all positive.

5. Experimental Results: Stereological-Type Data

The design of the experiment is shown in the block diagram in Figure 32.

Block 1. Data treatment — raw data from the Quantimet® 720 measurements was treated in the following manner:

1. It was assumed that any ten projections taken from a sample of particles of the same powder can be treated as representing one three-dimensional particle.
2. The mean values of these measurements were substituted into Equations 33, 34 to obtain the five-shape features and the two-size features for ten representative three-dimensional particles. These results were used as the training sets of patterns. Figures 33, 34, 35 show distribution curves for the features named: **AR**, SF1 and SF4 for the four powders.

Block 2. Determination of decision function — the training patterns from Step 1 were put into an iterative scheme using the algorithm derived from Equation 42. Four decision functions were obtained in the case of the four powders as shown in Figure 24. The four decision functions were composed of different sets of features according to the following scheme:

- Set 1 — AR, SF1, SF2, SF3, SF4, r, L3
- Set 2 — AR, AF1, SF2, SF3, SF4
- Set 3 — AR, SF2, SF3, SF4
- Set 4 — SF, SF1, SF3, SF4

All but Set 1 are particle size independent.

* A criterion function may have the form J(w,x) = (w'x − w'x), for example, it is a minimum when condition of Equation 40 is satisfied.

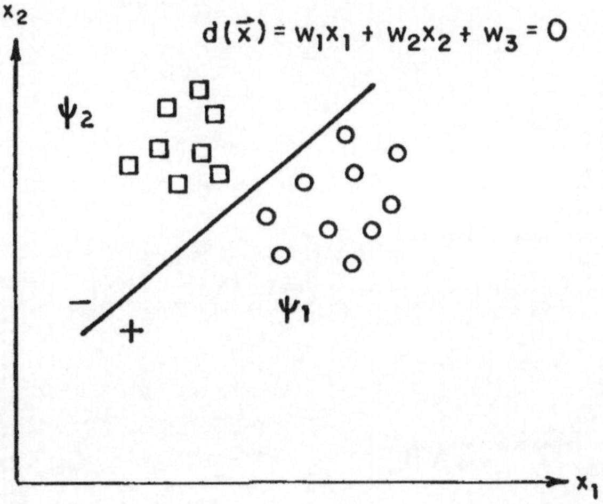

FIGURE 31. A simple decision function for two pattern classes. (From Beddow, J. K., *Powtech, 1977, Proceedings of the 4th International Poweder Technology and Bulk Solids Conference*, Heyden & Sons, London, 1977, 99. With permission.)

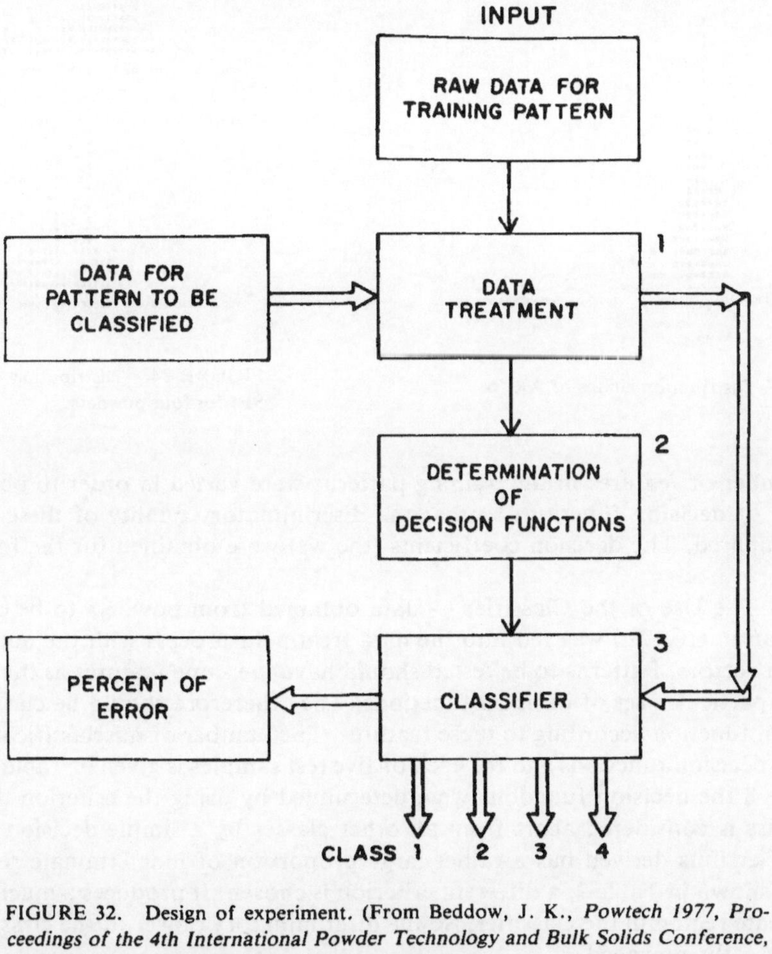

FIGURE 32. Design of experiment. (From Beddow, J. K., *Powtech 1977, Proceedings of the 4th International Powder Technology and Bulk Solids Conference*, Heyden & Sons, London, 1977, 99. With permission.)

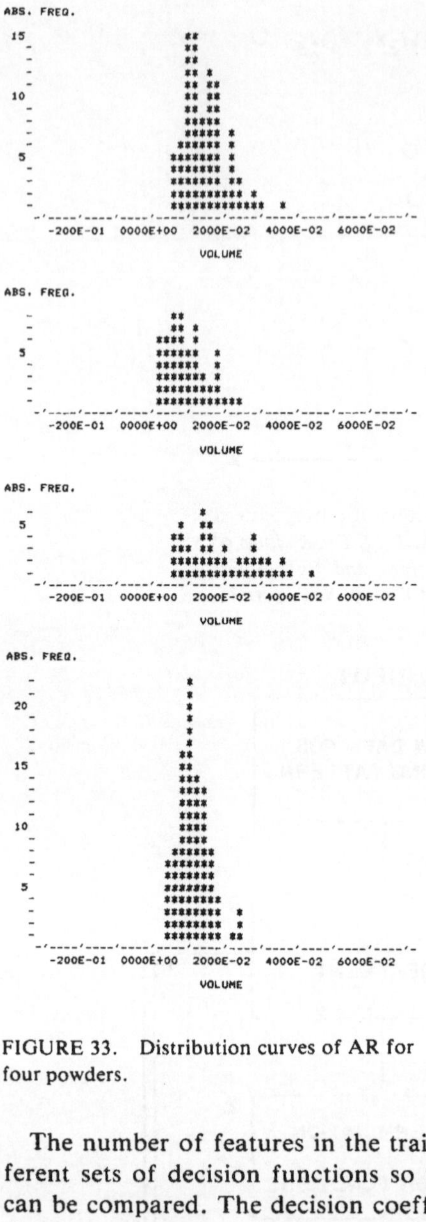

FIGURE 33. Distribution curves of AR for four powders.

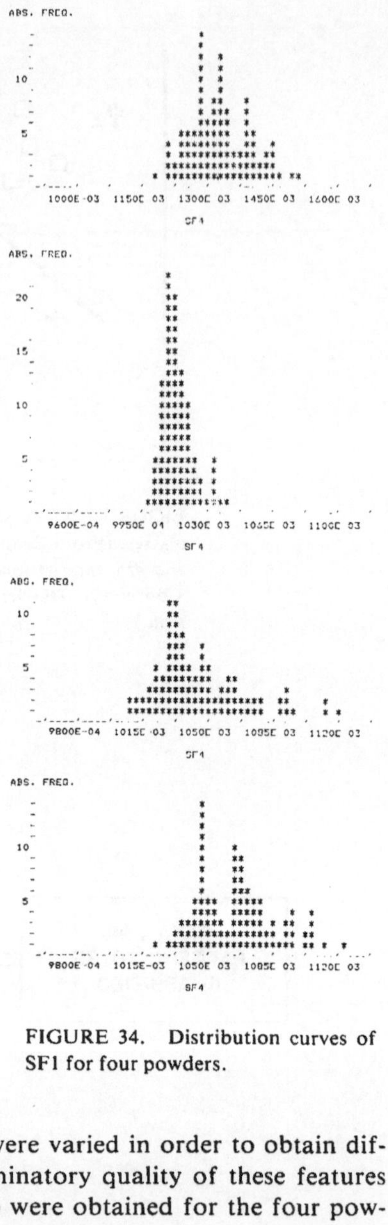

FIGURE 34. Distribution curves of SF1 for four powders.

The number of features in the training patterns were varied in order to obtain different sets of decision functions so that the discriminatory quality of these features can be compared. The decision coefficients (the ws) were obtained for the four powders.

Block 3. The Use of the Classifier — data obtained from powders to be classified by the Quantimet® 720 was fed into the data treatment process with the outcome of the pattern vectors. Patterns to be tested should have the same features as the training set in that particular set of decision functions. They therefore should be classified by the decision function according to these features. The number of misclassifications for each set of decision functions and for each of five test samples is given in Table 8.

In Table 8 the decision functions were determined by using the criterion that each pattern class is considered apart from all other classes by a simple decision surface. The classifier thus derived has a rather large proportion of indeterminate region. In the results shown in Table 9, a different criterion is chosen. It produces a much smaller indeterminate region in the classifier. So the discriminatory power of the classification scheme is greatly improved.

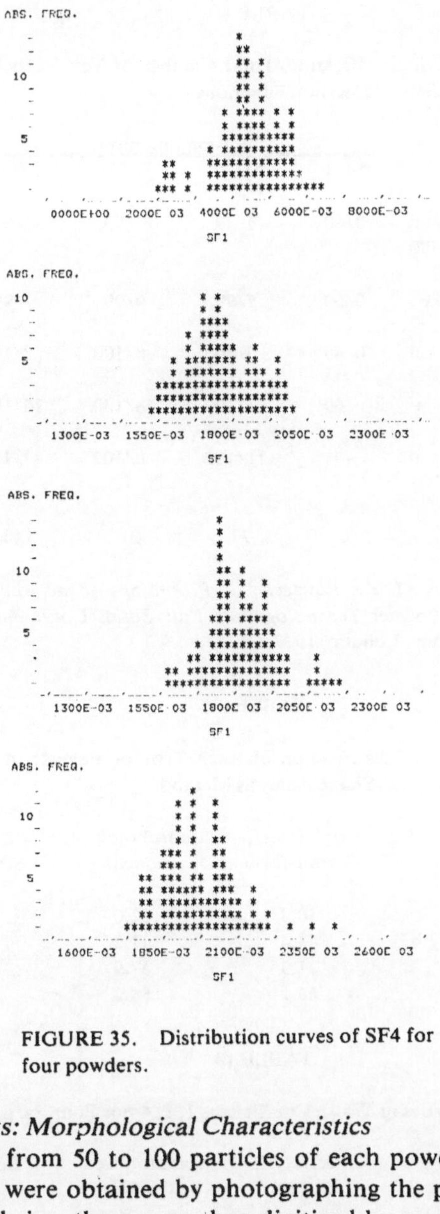

FIGURE 35. Distribution curves of SF4 for four powders.

6. Experimental Results: Morphological Characteristics

Sample size ranging from 50 to 100 particles of each powder set were selected and analyzed. The profiles were obtained by photographing the particle silhouettes. After enlarging to a standard size, these were then digitized by means of a two-dimensional Grafpen® system. This reports the profile as a set of (x,y) coordinates which are then processed to obtain the set of Walsh coefficients required.

The following results are reported: the statistical properties of three Walsh coefficients for each of the four particle samples. These are given in Table 10. The means and the standard deviations for each morphological feature for each of the four particle samples are given in Table 11. Examples of profiles of low-, medium-, and high-valued coefficients for each of four particle samples are given in Figure 36. Correlations for the morphological features are reported in Table 12. Distributions curves for the two Walsh coefficients are given in Figures 37 and 38 and for volume and corner in Figures 39 and 40. The number of misclassifications for the morphological features selected for each particle set of two are given in Table 13. These may be compared with misclassifications for Walsh and stereological-like features.

TABLE 8

Number of Misclassification/Total Number of Test Particles for the Four Sets of Decision Functions

Class	Decision Functions			
	Set 1	Set 2	Set 3	Set 4
Gas-atomized Cu-PB + 170	0/60	48/49	54/99	30/99
Electrolytic Fe + 170	0/60	7/99	9/99	7/99
Electrolytic Cu + 170	0/60	0/100	0/100	8/100
Sponge Fe + 170	0/60	11/100	28/100	20/100
Gas-atomized Cu-Pb 270/325	—	11/102	0/102	83/102
Σ (errors)		**77**	**91**	**148**

From Beddow, J. K., *Powtech '77, Procedings of the 4th International Powder Technology and Bulk Solids Conference*, Heyden & Sons, London, 1977, 99.

TABLE 9

Percentage Error in Classification of Each Type of Particle in a Four-Class Classifier for Each Shape Analysis Method

	R transform	Morphological modeling	Stereology
Electrolytic Cu	30.3	82.0	0.0
Gas-atomized Cu-Pb	43.4	47.2	1.0
Electrolytic Fe	31.3	47.9	0.0
Sponge Fe	63.5	58.3	8.0

TABLE 10

Statistical Properties of Walsh Coefficients 1, 3, 4 For Four Particle Sets

Walsh Coefficient		Electrolytic Cu (89)	Gas atomized Cu-Pb	Sponge Fe	Electrolytic Fe
Coefficient 1	minimum	1.0720	1.4885	1.4574	1.3579
	maximum	2.2941	2.2394	3.0037	2.4436
	mean	1.5356	1.7717	2.2062	1.8360
	std deviation	0.2485	0.1870	0.3179	0.2333
Coefficient 3	minimum	0.0299	0.0041	0.0258	0.0169
	maximum	0.2385	0.2310	0.2690	0.2344
	mean	0.0972	0.0553	0.0944	0.0775
	std deviation	0.0415	0.0419	0.0505	0.0403
Coefficient 4	minimum	0.0709	0.0253	0.0793	0.0236
	maximum	0.6605	0.8778	1.2710	0.7223
	mean	0.2798	0.2769	0.5601	0.2693
	std deviation	0.1397	0.2070	0.3138	0.1437

From Lew, G., Beddow, J. K., and Vetter, A. F., *Powder Metallurgy Int.*, submitted, 1979. With permission.

TABLE 11

Statistical Properties of Named Morphological Characteristics For Four Particle Sets

		Electrolytic Cu sample size = 89	Gas atomized Cu-Pb sample size = 53	Sponge Fe sample size = 96	Electrolytic Fe sample size = 48
Arc	minimum	89.88	27.66	75.74	88.93
Length	maximum	753.71	343.87	506.36	648.71
	mean	288.18	140.29	179.33	276.63
	std deviation	135.85	76.83	83.47	151.85
Side	minimum	117.13	26.19	91.75	109.02
Area	maximum	1056.19	478.60	701.31	908.41
	mean	397.15	185.16	240.01	375.91
	std deviation	193.57	109.77	118.28	214.95
Verticle	minim	790.06	186.72	579.26	668.99
Edge	maximum	9344.67	3360.40	5658.77	7174.99
	mean	3125.05	1231.59	1694.85	2546.89
	std deviation	1733.43	787.69	970.83	1661.30
Positive	minimum	5.61	1.28	3.83	5.43
Block	maximum	35.31	24.02	24.53	43.25
Volume	mean	14.90	9.65	11.17	19.21
	std deviation	5.82	5.61	4.29	9.91

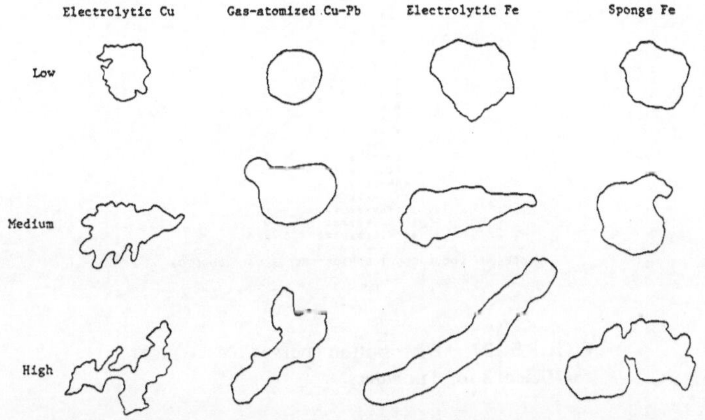

FIGURE 36. Examples of profiles of low, medium, and high morphological characteristic. (From Lew, G., Beddow, J. K., and Vetter, A. F., *Powder Metallurgdy Int.*, submitted, 1979. With permission.)

7. *Discussion: Stereological Features*

The work reported above was predicated on the hypothesis that the methods of pattern recognition can be applied to the problems of particle shape classification. The results obtained support this hypothesis.

It has been demonstrated that it is possible to construct a set of shape features and to design a classifier and use the features as pattern vectors.

Different sets of decision functions were used in the case of four metal powders. Each with a different degree of success vis-à-vis the error associated with the classification attempts made with them.

The original design of the classifier considered each pattern class as being apart from all other classes by a simple decision surface. This produced the errors shown in Table

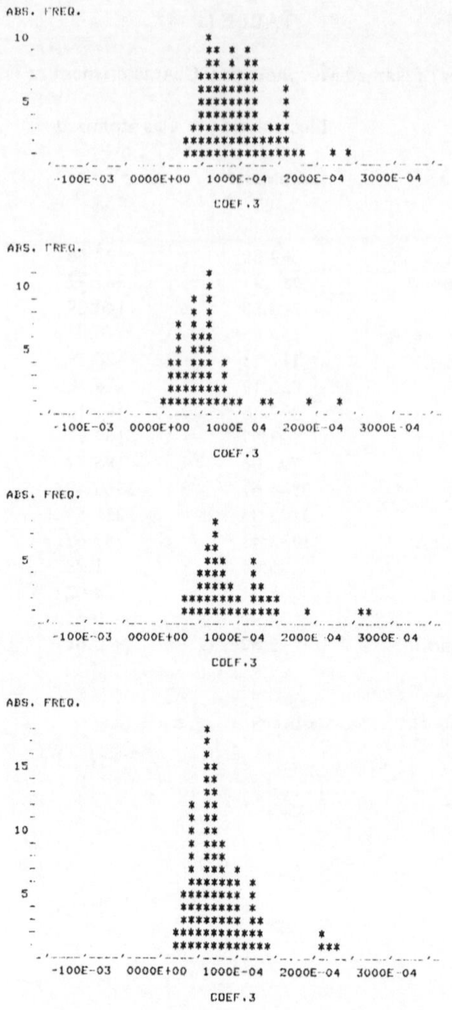

FIGURE 37. Distribution curves of Walsh coefficient 3 for 4 powders.

8. A redesign of the classifier produced sharply reduced classification errors for the same populations of particles as is shown in Table 9. (The classifier was modified to a pair-wise classification and certain subroutines were improved also.

8. Discussion: Morphological Characteristics

Examination of the distribution curves of the morphological characteristics shows that most appear to be skewed and generally towards higher values. The tables of correlations show high correlations between most if not all morphological coefficients. While this is not surprising, because they are all functions of n and An, it does pose something of a problem with respect to the design of a classifier. However, indications are that a high degree of discrimination is possible between the sets of the morphological characteristics from the different powders. (See section on Statistical Classifiers).

Figure 36 illustrates examples of profiles of low, medium, and high morphological coefficients (read from top to bottom) for the four powders studied. As is to be expected the direction of increased value of the coefficient corresponds to an increased amount of information in the profile. Examination of Figure 36 shows that the elecro-

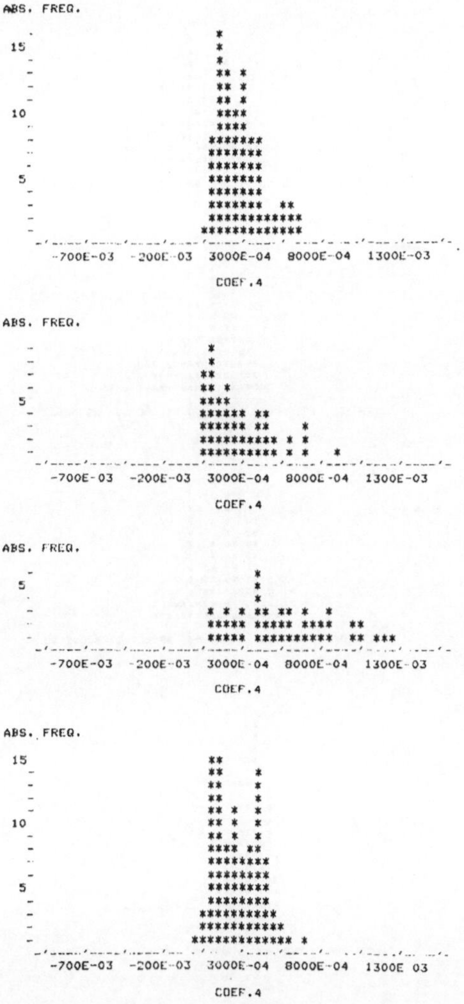

FIGURE 38. Distribution curves of Walsh coefficient 4 for 4 powders.

lytic copper particles in particular are not holomorphic. They have reentrants. The method used here, of transforming the particle profile into a Walsh function is unsuitable for such nonholomorphic particles. In practice this means that for this part of the data set we do not know what information we have lost, if any.

In general the morphological features appear to be unsuitable as features in a pattern classifier. This is especially obvious when the misclassifications are compared with those obtained using stereological properties. However, this is not quite a clear-cut argument for the following reasons:

(1.) It will be shown later (section on Statistical Classifiers) that fairly reasonable (i.e., low) misclassification scores can be obtained if the statistical properties of the features are taken into account.

(2.) The morphic features have skewed distributions and large standard deviations as compared with the stereological features (Table 11 and Figures 37, 38, 39 and 40). One would expect to get lower classification success with the former as compared with the latter.

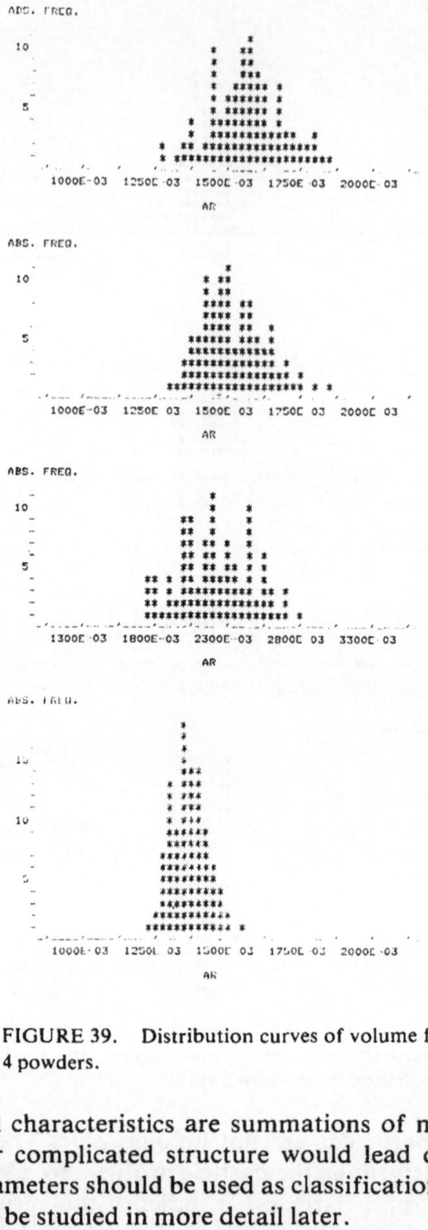

FIGURE 39. Distribution curves of volume for 4 powders.

(3.) The morphological characteristics are summations of many coefficients to various powers. Their complicated structure would lead one to expect that more simply derived parameters should be used as classification features initially. Items of interest can then be studied in more detail later.

(4.) Even using the simpler Walsh coefficients themselves does not appear to have reduced the misclassification to an acceptable level in the deterministic classifier. This indicates the necessity of a feature selection step, which is currently being developed.

(5.) Because the (R,Θ) method is unsuitable for reentrant shapes, it is difficult to make a clear-cut decision on the data for the electrolytic copper in particular. The (ϕ,ℓ) program will be used in the near future to study the data further.

(6.) The data massaging in the case of the stereological analysis will tend to produce normally distributed input data to the classifier. One would naturally expect better results with this sort of input, in terms of low misclassification scores, as compared with skewed input data.

(7.) As shown in Table 13, features from R transform produces a better classifier than those from morphological modeling. Since only the first 6 coefficients out

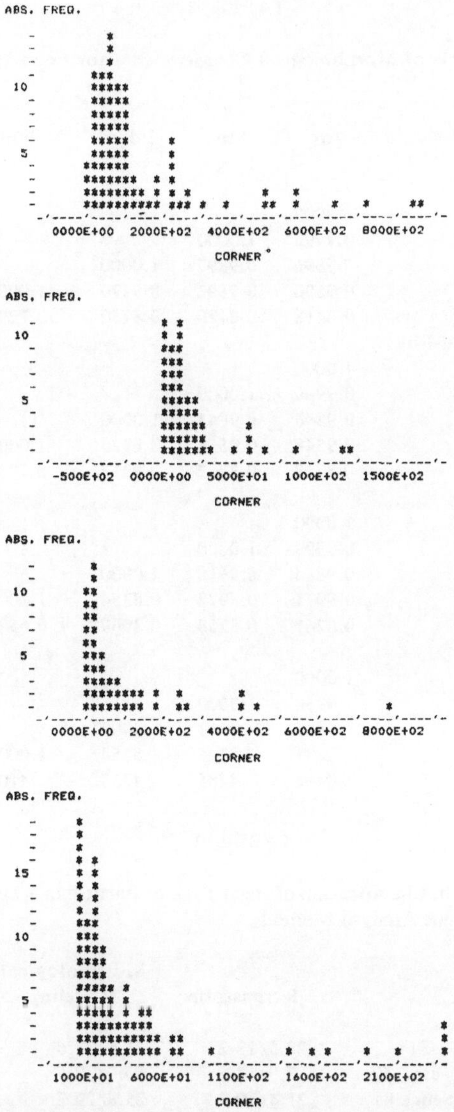

FIGURE 40. Distribution curves of corner for 4 powders.

of 128 were looked into, a lot of information was skipped. Przygocki[10] has shown that feature of greater discriminatory power can usually be found if more coefficients are examined. He obtained his result from Fourier coefficients. However, this should also be true with Walsh coefficients.

D. Statistical Classifiers[18]

This study presents pair-wise classifications of four sets of metal powders via shape analysis. The four types of powders (and their sample sizes) are: gas atomized Cu-Pb(53), electrolytic Cu(89), electrolytic Fe(48), sponge Fe(96). In this study we used Walsh coefficients as classification features in a number of statistical classifiers. The R transform was used to generate the Walsh coefficients. The (R,Θ) method of analysis was used.

TABLE 12

Correlation Table of Morphological Characteristics for Four Types of Particles

	Arc	Area	Edge	Volume	Corner
Electrolytic Cu					
Arc	1.0000				
Area	0.9999	1.0000			
Edge	0.9896	0.9897	1.0000		
Volume	0.9596	0.9592	0.9170	1.0000	
Corner	0.8418	0.8420	0.8750	0.7299	1.0000
Gas-atomized Cu-Pb					
Arc	1.0000				
Area	0.9999	1.0000			
Edge	0.9846	0.9843	1.0000		
Volume	0.9548	0.9554	0.8925	1.0000	
Corner	0.8728	0.8712	0.9313	0.7281	1.000
Electrolytic Fe					
Arc	1.0000				
Area	0.9999	1.0000			
Edge	0.9814	0.9618	1.0000		
Volume	0.8971	0.8972	0.8354	1.0000	
Corner	0.8267	0.8258	0.8969	0.6524	1.0000
Sponge Fe					
Arc	1.0000				
Area	9958	1.0000			
Edge	9875	9913	1.0000		
Volume	9281	9325	8354	1.0000	
Corner	4488	4485	4733	4101	1.0000

TABLE 13

Percentage Error in Classification of Each Type of Particle in a Two-Class Classifier for Each Shape Analysis Method

	R transform	Morphological Modeling	Stereology
Electrolytic Cu/Cu-Pb	22.5/13.2	21.5/22.6	0.0/0.0
Electrolytic Cu/electro. Fe	7.9/12.5	60.7/37.5	0.0/0.0
Electrolytic Cu/sponge Fe	21.4/30.2	25.8/29.2	0.0/0.0
Cu-Pb/electrolytic Fe	18.9/18.8	24.5/39.6	0.0/2.0
Cu-Pb/sponge Fe	30.0/33 .0	47.2/29.2	0.0/7.0
Electrolytic Fe/sponge Fe	27.0/18.8	41.7/28.1	0.0/0.0

1. Feature Selection

The profiles of the particles are approximated by the first 64 terms of Walsh functions.[5] Since direct approximation by Walsh functions will cause the coefficients being translation variant, we need to use the R transform[8] of Walsh functions to provide the translation-invariant property. However, detailed information about the shapes of the particles is lost when the R transform is used. Figure 41 shows the means of the first 64 transformed Walsh coefficients for 15 Cu-Pb and electrolytic Fe samples. It shows that the means of the coefficients for the two types of powders approach zero as the order of terms increases. Therefore, only the first few terms may be considered as classification features. By examining the distribution of the first ten coefficients for various powders, it was decided that the first three coefficients be used as the classification features. The other coefficients either have large variances or have little discrim-

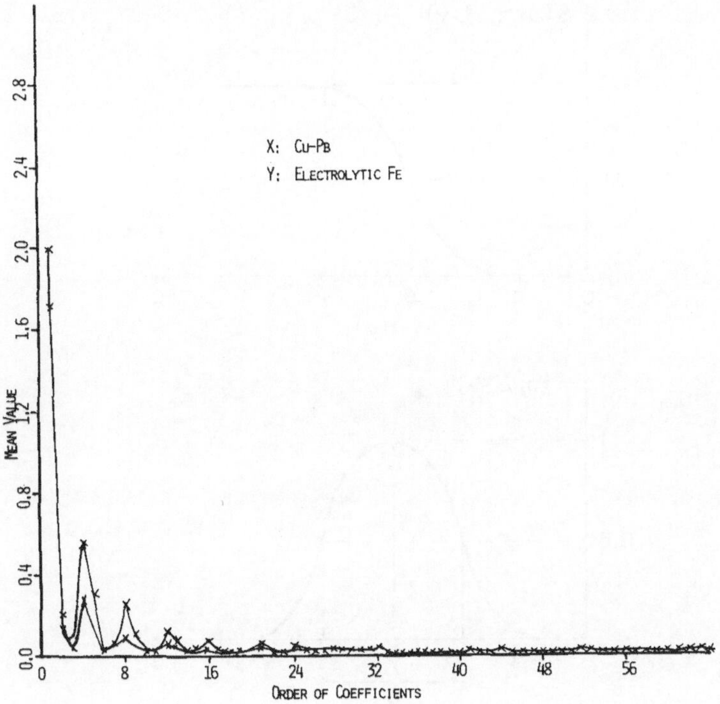

FIGURE 41. Plot of means for the Walsh coefficients. (From Wang, C., Beddow, J. K., Vetter, A. F., and Lew G., submitted for publication, 1978.)

inatory power. It was hoped that those three coefficients would provide enough information about the shapes of the particles for reasonable classification.

2. Classification Rules

There are three basic classification rules used in the study: linear discriminant function (LDF) rule,[26] quadratic discriminant function (QDF) rule,[26] and Fix-Hodges k-nearest neighbor rule.[28] The LDF and QDF rules are derived from the assumption that the underlying distributions of the populations being multivariate normals, while the Fix-Hodges rule is a nonparametric decision rule (i.e., a decision rule without assuming that the underlying distributions are known). Let $x_1, \ldots, x_n$ and $y_1, \ldots, y_n$ be the samples from populations that have equal prior probabilities of occurrence and equal costs of misclassification. A brief description of each of these three procedures follows.

3. Linear Discriminant Function (LDF) Rule

Suppose populations π_1 and π_2 are normally distributed with means $\underset{\sim}{\mu}_1$, $\underset{\sim}{\mu}_2$, and common covariance matrix $\underset{\sim}{\Sigma}$. From Bays' formula, the optimal decision rule which minimizes the expected total probability of misclassification is: for an observation $\underset{\sim}{Z}$, if $D_L(\underset{\sim}{Z}) > 0$, then classify $\underset{\sim}{Z}$ into π_1: otherwise classify $\underset{\sim}{Z}$ into π_2, where

$$D_L(\underset{\sim}{Z}) = [\underset{\sim}{Z} - \frac{1}{2}(\underset{\sim}{\mu}_1 + \underset{\sim}{\mu}_2)]'\Sigma^{-1}[\underset{\sim}{\mu}_1 - \underset{\sim}{\mu}_2] \quad (43)$$

If the populations parameters μ_1, μ_2, and Σ are unknown, the discriminant function $D_L(\underset{\sim}{Z})$ may be approximated by

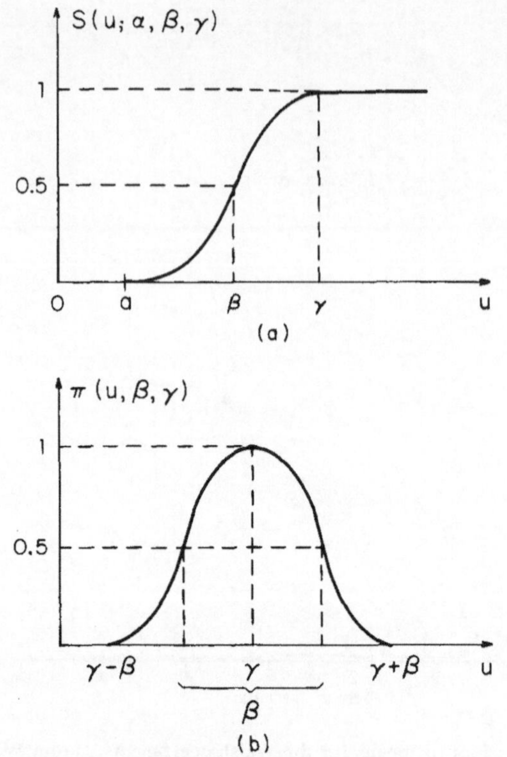

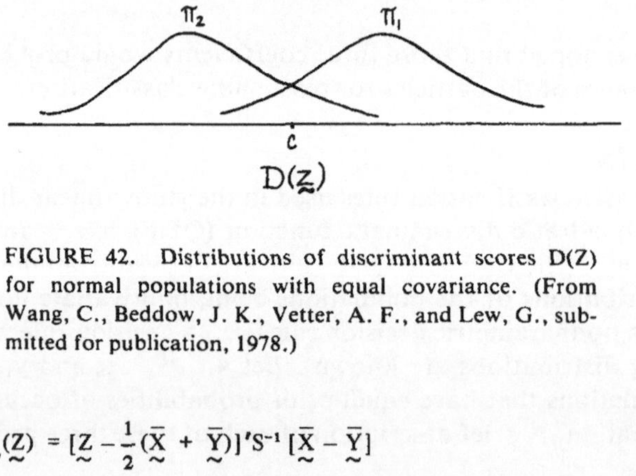

FIGURE 42. Distributions of discriminant scores D(Z) for normal populations with equal covariance. (From Wang, C., Beddow, J. K., Vetter, A. F., and Lew, G., submitted for publication, 1978.)

$$D_L(\underset{\sim}{Z}) = [\underset{\sim}{Z} - \frac{1}{2}(\underset{\sim}{\bar{X}} + \underset{\sim}{\bar{Y}})]'S^{-1}[\underset{\sim}{\bar{X}} - \underset{\sim}{\bar{Y}}]$$

$$S = [(m-1)S_1 + (n-1)\underset{\sim}{S_2}]/(m+n-2) \qquad (44)$$

The distributions of the discriminant scores ($D_L(.)$ are shown in Figure 42. The point C is the cut-off point such that an observation Z is classified into π_1 if $D_l(Z) > C$, otherwise classify Z into π_2.

The LDF rule has been one of the most commonly used classification rules because of its robustness and ease of computation. Some previous applications with the use of the LDF rule are, for example, crop classification[29] and diagnosis of black lung.[30] The performance of the LDF rule has been studied extensively in a number of nonoptimal conditions (i.e., the underlying distributions of the populations are nonnormal and/or have unequal covariance matrices), for example, by Gilbert,[31] and Lachenbruch, et al.[32] In general, the LDF rule has been found to be quite robust. In this study, the distributions of the samples are mostly skew to the right, but not very far from normal. Figure 37 shows the sample distributions of the third Walsh coefficient for Cu-Pb and electrolytic Fe samples. The distributions of other coefficients are similar to the distributions shown in Figure 37. The LDF rule should perform fairly well if the covariance matrices of the two populations being considered are relatively the same.

4. *Quadratic Discriminant Function (QDF) Rule*

Suppose populations π_1 and π_2 are normally distributed with means $\underset{\sim}{\mu}_1$ and $\underset{\sim}{\mu}_2$, covariance matrices Σ. From Baye's formula, the optimal decision rule which minimizes the expected total probability of misclassifications is: for an observation $\underset{\sim}{Z}$, if $D_Q(\underset{\sim}{Z}) > 0$, then classify $\underset{\sim}{Z}$ into π_1; otherwise classify $\underset{\sim}{Z}$ into π_2, where

$$D_Q(\underset{\sim}{Z}) = (\underset{\sim}{Z} - \underset{\sim}{\mu}_2)'\Sigma_2^{-1}(\underset{\sim}{Z} - \underset{\sim}{\mu}_2) - (\underset{\sim}{Z} - \underset{\sim}{\mu}_1)'\Sigma_1^{-1}(\underset{\sim}{Z} - \underset{\sim}{\mu}_1) + \ln(|\Sigma_2|/|\Sigma_1|) \tag{45}$$

If the populations parameters are unknown, the discriminant function $D_Q(\underset{\sim}{Z})$ may be approximated by

$$D_Q(\underset{\sim}{Z}) = (\underset{\sim}{Z} - \bar{\underset{\sim}{Y}})'S_2^{-1}(\underset{\sim}{Z} - \bar{\underset{\sim}{Y}}) - (\underset{\sim}{Z} - \bar{\underset{\sim}{X}})'S_1^{-1}(\underset{\sim}{Z} - \bar{\underset{\sim}{X}}) + \ln(|S_2|/|S_1|) \tag{46}$$

The QDF rule also has been used frequently in practice. The use of QDF rule is recommended when it is suspected that the covariance matrices of the two populations are not equal. In this study, the assumption of unequal covariances is sometimes a more realistic one. The performance of the QDF rule would be better than that of the LDF rule in those cases.

5. *Fix-Hodges k-Nearest Neighbor Rule*

Let $\underset{\sim}{Z}_1, \ldots, \underset{\sim}{Z}_k$ be the k-nearest points to an observation Z in terms of some distance measure. If n_1 and n_2 are the number of $\underset{\sim}{Z}_1, \ldots, \underset{\sim}{Z}_k$ from population π_1 and π_2 ($n_1 + n_2 = k$), respectively, then the k-nearest neighbor rule classifies $\underset{\sim}{Z}$ into π_1 if $n_1 > n_2$; otherwise, classifies $\underset{\sim}{Z}$ into π_2. In this study, the value of k is set to 7 and the Euclidian distance measure is used.

The essential idea of this classification rule is that we attempt to classify $\underset{\sim}{Z}$ into the population which has the larger concentration of probability in a region around $\underset{\sim}{Z}$. The near optimal performance and simplicity makes it an attractive decision rule.

In addition to the regular LDF and QDF rules, the rank procedures of LDF and QDF are also considered. A rank procedure[33] tends to balance the probabilities of misclassification between the two populations, while maintaining the classification capability of a given discriminant function. The development of a rank procedure from a discriminant function D(.) proceeds as follows:

1. Given an observation $\underset{\sim}{Z}$, first assume that $\underset{\sim}{Z}$ belongs to π_1. Based on the samples $\underset{\sim}{Z}, \underset{\sim}{X}_1, \ldots, \underset{\sim}{X}_m$ and $\underset{\sim}{Y}_1, \ldots, \underset{\sim}{Y}_n$, a discriminant function $D_X(.)$ is computed.
2. Compute the discriminant scores $D_X(Z)$, $D_X(X_1), \ldots, D_X(X_m)$, and $D_X(\underset{\sim}{Y}_1), \ldots, D_X(\underset{\sim}{Y}_n)$. Let $R_X(Z)$ be the rank of $D_X(Z)$ among $D_X(\underset{\sim}{Z})$, $D_X(\underset{\sim}{X}_1), \ldots, D_X(\underset{\sim}{X}_m)$, ranking from smallest to largest.

3. Assume that $\underset{\sim}{Z}$ belongs to π_2. Based on $\underset{\sim}{X}_1, \ldots, \underset{\sim}{X}_m$, and $\underset{\sim}{Z}, \underset{\sim}{Y}_1, \ldots, \underset{\sim}{Y}_n$, a discriminant function $D_Y(.)$ is computed.
4. Compute $D_Y(\underset{\sim}{X}_1), \ldots, D_Y(\underset{\sim}{X}_m)$, and $D_Y(\underset{\sim}{Z}), D_Y(\underset{\sim}{Y}_1), \ldots, D_Y(\underset{\sim}{Y}_n)$. Let $R_Y(\underset{\sim}{Z})$ be the rank of $-D_Y(\underset{\sim}{Z})$ among $-D_Y(\underset{\sim}{Z}), -D_Y(\underset{\sim}{Y}_1), \ldots, -D_Y(\underset{\sim}{Y}_n)$.
5. Let $P_X(\underset{\sim}{Z}) = R_X(\underset{\sim}{Z})/(m + 1)$ and $P_Y(\underset{\sim}{Z}) = R_Y(\underset{\sim}{Z})/(n + 1)$. The rank procedure is then of the form:
   (i) If $P_X(\underset{\sim}{Z}) > P_Y(\underset{\sim}{Z})$, classify $\underset{\sim}{Z}$ in π_1
   (ii) If $P_X(\underset{\sim}{Z}) < P_Y(\underset{\sim}{Z})$, classify $\underset{\sim}{Z}$ in π_2
   (iii) If $P_X(\underset{\sim}{Z}) = P_Y(\underset{\sim}{Z})$, classify $\underset{\sim}{Z}$ using a nonrank procedure

In this study, the regular LDF or QDF rule is used in case of tie.

It is sometimes desirable to rank the combined original data from both populations componentwise and then use the ranks rather than the original data in the construction of classification rules. This would alleviate the sensitivity of the data when the original data has a large variance. In this study, both the original data and the rank data are used for the LDF and QDF rules. Table 14 summarizes the classification rules and data used for this study.

6. Classification Results and Discussion

Pair-wise classification for the four sets of particles are made using the classification rules discussed above. The estimated probabilities of misclassification for various combinations are given in Table 15. The U-method suggested by Lachenbruch and Mickey[34] has been used to estimate the probabilities of misclassification.

The results show that the degree of success depends on the particular pair of powders used, although in general the results are quite satisfactory. Also notice that no single classification rule is consistently the best for all cases. This suggests further development of other classification rules to be considered for particle classification.

The results presented in this section indicate that a satisfactory way of classifying particle profiles on a basis of their Walsh coefficients could be developed. Enough has been learned to give some pointers which will assist this development:

1. Although the R transform smears the higher-order Walsh coefficients, the lower-order coefficients are quite satisfactory for use in modeling of particle features, as has been described above. In this study we have used morphic descriptors as the features with which to discriminate members of one set of particles from another set. There is, however, an alternative viewpoint: the main aim of discriminant analysis is to do just that — i.e., discriminate. An alternative strategy therefore is to utilize the set of sets of descriptors as features which give the lowest misclassification score. Following this step, further analysis can alwas be conducted upon the particle or particles of concern. This appears to be one basic idea behind the work of Ehrlich.
2. The ultimate aim of discriminant analysis being to separate into viable meaningful classes; as we learn more about morphological analysis it should be possible to develop decision rules that will enable the analysis to separate data from the main stream where this is appropriate, so that the remaining data will be more refined, concentrated, and lead to lower misclassification scores. The idea behind this strategy is to isolate unusual particles so that they do not overload the discriminatory capability of the system being used to analyze the data.
3. As the ultimate aim of morphological analysis is to relate morphic characteristics to mode of origin and mode of behavior, one should strive to utilize classification features that permit one to model mechanisms of formation and behavior, respectively.

TABLE 14

Summary of Procedures

Procedure Number	Description
1	Nonrank procedure, LDF, original data
2	Nonrank procedure, QDF, original data
3	Nonrank procedure, Fix-Hodges, original data
4	Rank procedure, LDF, original data
5	Rank procedure, QDF, original data
6	Nonrank procedure, LDF, rank data
7	Nonrank procedure, QDF, rank data
8	Rank procedure, LDF, rank data
9	Rank procedure, QDF, rank data

From Wang, C., Beddow, J. K., Vetter, A. F., and Lew, G., *Proceedings: International Powder and Bulk Solids Handling and Processing Conference,* Industrial and Scientific Conference Management, Chicago, 1978, 158. With permission.

TABLE 15

Percentages of Misclassification

Procedure	Cu-Pb	Electrolytic Cu	Average	Procedure	Cu-Pb	Electrolyic Fe	Average
Population I: Cu-Pb Population II: Electrolytic Cu				**Population I: Cu-Pb Population II: Electrolytic Fe**			
1	9	24	18	1	13	23	18
2	13	28	23	2	19	23	21
3	9	21	17	3	19	23	21
4	15	17	16	4	17	19	18
5	17	20	19	5	23	21	22
6	13	20	18	6	17	19	18
7	15	21	19	7	21	21	21
8	15	17	16	8	17	19	18
9	15	17	16	9	21	21	21
Population I: Electrolytic Cu Population II: Sponge Fe				**Population I: Electrolytic Fe Population II: Sponge Fe**			
1	28	45	39	1	8	17	11
2	28	30	30	2	9	15	11
3	32	33	33	3	9	10	9
4	34	34	34	4	11	13	12
5	26	28	28	5	11	10	11
6	34	38	36	6	17	6	13
7	30	32	32	7	15	8	12
8	36	36	36	8	11	10	11
9	32	32	32	9	11	10	11
Population I: Cu-Pb Population II: Sponge Fe				**Population I: Electrolytic Cu Population II: Electrolytic Fe**			
1	21	18	19	1	27	24	25
2	25	17	21	2	35	18	24
3	26	17	21	3	29	19	22
4	19	20	19	4	27	26	26
5	21	22	22	5	25	26	26
6	22	18	20	6	23	29	27
7	22	16	19	7	23	29	27
8	20	20	20	8	25	26	26
9	20	22	21	9	27	27	27

From Wang, C., Beddow, J. K., Vetter, A. F., and Lew, G., *Statistical Classifiers in Particle Morphology Analysis,* submitted for publication.

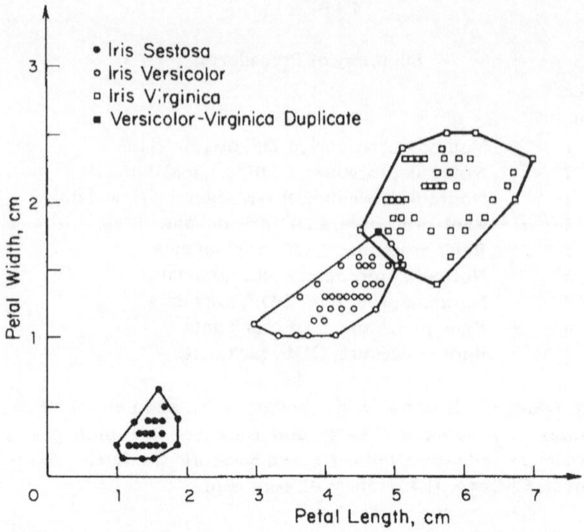

FIGURE 43. Anderson's Iris data-petal features. (From Bezdek, J., *J. Math. Biol.*, 1(1), 57, 1973. With permission.)

4. Keeping in mind the analog of chemical analysis for a moment, it should not be surprising that many different techniques will be necessary in order to conduct morphological analysis on finely divided matter. This therefore parallels chemical analysis wherein there are many techniques, each used where it is most appropriately employed.

E. Fuzzy Classifiers[21]

The applicability of fuzzy clustering methods to biological sciences has been underscored by the comment due to Adey[35]; "Our capability to interpret EEG patterns in cerebral systems has been sustantially enhanced by pattern recognition techniques, principally stepwise discriminant analysis, and more recently, by fuzzy clustering techniques."

A basic fuzzy clustering technique due to Bezkek has been introduced in Equation 28. The Iris data due to Anderson is a well-known set of four dimensional botanical data. Figure 43[23] depicts the data plotted for two dimensions: petal length and petal width. These are divided into three convex hulls as shown. By using the fuzzy algorithm, and without prior knowledge of the labeled pattern, a solution depicted in Figure 44[23] was obtained. It was concluded that the technique of fuzzy clustering provides a reliable criterion for posteriori evaluation of cluster validity. The potential of this technique to the morphological analysis of fine particles has yet to be attempted. The method is being used, however, in Fourier harmonic analysis of sediment mixtures.[9]

VII. VERBAL AND GEOMETRIC DESCRIPTORS

Pattern recognition instruments are expensive and therefore they are, more often than not, unavailable. They are normally used as a service in a research and development organization. Human observers are therefore forced to rely on nonmathematical descriptors — mainly geometrical and verbal descriptors, which are totally unorganized in the sense that they follow no logical mathematical framework. There seems to be a need, therefore, for the development of another framework as a part of a three-step system for particle shape analysis.

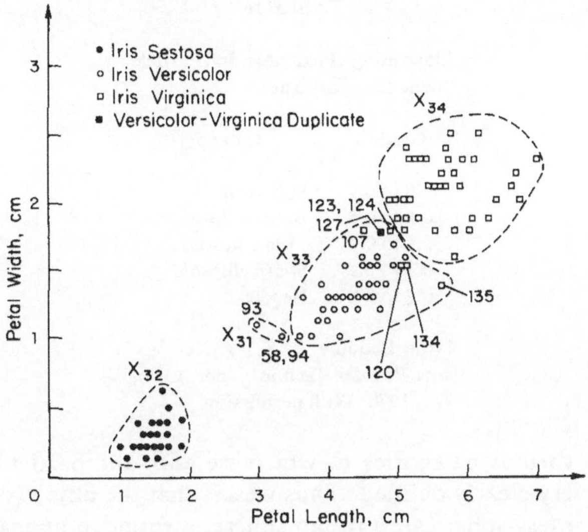

FIGURE 44. Sequential solution via fuzzy ISODATA. (From Bezdek, J., *J. Math. Biol.*, 1(1), 57, 1973. With permission.)

This proposed systematic procedure for analyzing the shapes of individual particles provides, for powder processing and research alike, a range of description from the gross to the minute which will meet the needs in the complete spectrum from low-resolution analysis for purposes of general communication to observation, selection, and measurement:

Step 1: comprises a heuristic shape comparator subsystem, the use of which will provide the observer with an overview of the shape characteristics of the sample;

Step 2: is a judgment step in which the observer decides whether to seek additional information, and, if so, what method(s) to select for use at what cost; and

Step 3: requires the observer to apply one method (or more) from among the four classes of shape analysis methods correctly available, to measure the points and hence determine shape.

Step 1 will provide shape description commonly required by most general users. Each successive step will provide additional information of differing complexity and all data will fit into the prescribed system of description and notation so that users can communicate precisely with each other.

We have opted for a heuristic Step 1 on the grounds of flexibility. In addition we have laid out the requirements that the user of Step 1 should be involved in as little work as possible in using it, a minimum of equipment should be involved, and it is clearly desirable that the method be not inconsistent with further theoretical development, and finally Step 1 should not require a high degree of special skill for successful usage.[36]

When an observer examines the shape of an object, the rapid eye scanning that occurs in no way attempts to account for all the points of surface, but some sort of sampling operation goes on. If the observer is very familiar with the shape being observed, the scan is quickly completed and the classifying decision almost simultaneously made. The interesting point here is the first thing that we do when we attempt to decide on the shape of a particular object is use our built-in comparator based system of classification, with the categories structured by our experiences and based upon some logical process, presumably.

It is also interesting and quite important to note that when we carry out our classi-

TABLE 16

Illustrating Fuzziness in Particle Shape Categorization

Grade	Category
1.0 to 0.9	Spherical
0.9 to 0.8	Bulky ellipsoid
0.8 to 0.4	Thin ellipsoid
0.4 to 0.2	Sharp ellipsoid
0.2 to 0.0	Needle

From Beddow, J. K., *J. Res. Assoc. Powder Technol. Jpn.*, 15(4), 227, 1978. With permission.

fying activity, the various categories to which we allot the particular object's shape are not exactly or very clearly defined. Thus we say that the object's shape is rounded, spheroidal, more or less spherical, a rough sphere, a rounded appearance, and so on. All of these descriptions convey the idea of sphericity, without stating that the object is a sphere. And, again, what may be rounded to one may be described by another as spheroidal. Both observers may each quite clearly understand the other with little difficulty, or they may not to an extent which in some part must depend upon the commonality of experience of the observers.

In the case of particles produced from liquids, minimization of surface energy is of importance in the shape formation,[37] and, again, in the case of particles produced by comminution, a minimum energy of fracture path is of importance[38] in the shape formation. Clearly, the former produces rounded shapes and the latter can produce angular shapes. A geometrical basis would be a generalization sufficiently broad to cover both examples; it would also possess the capability of being specific to each, always provided that we are able to develop a satisfactory procedure for developing shape categories.

The shape of a particle is an intrinsic characteristic; it represents the quality of a portion of matter. In this sense, therefore, to compare the shape of one particle with that of another is to compare their qualities with reference to some specific characteristic(s). This comparison procedure may be greatly facilitated by applying the concept of the fuzzy set.

In mathematics, a set is usually conceived as having a precisely defined boundary such that one can distinguish readily between those members that belong and those that do not. In an ordinary set, therefore, there are two levels of belonging represented by 0 for a nonmember and by 1 for a member.

By way of contrast in the real world sets do not have sharp boundaries. It is often extremely difficult to decide whether a particular object is a member of a set or not. In response to this problem, the concept of the fuzzy set has been introduced.[39] A fuzzy set is a class in which there is graded membership. Thus, all members have in common only that their membership function is positive (i.e., lying somewhere between zero and one), but their grade of membership runs the gamut between zero and one. The relevance of the fuzzy set to the problem in hand is shown in the following example: consider a group of shapes consisting of a sphere and a variety of ellipses of gradually increasing eccentricity. Let us arbitrarily choose five specific shapes, one sphere and four ellipsoids. If we take the ratio of the minor to major axis as the membership criterion one could state the grades of membership as shown in Table 16.

In a fuzzy set, the source of fuzziness may be either that the goals are not sharply defined or that the constraints are not sharply defined (or both are not sharply defined).[40] Statements illustrating fuzziness in respect of particle shape include: the particles are flaky; the particles are rounded, etc. The meaning of flaky and rounded is not sharply defined. When either the goals or the constraints of both constitute classes of alternatives whose boundaries are not sharply defined the decision process (e.g., the judgment as to the shape of the particle) is an example of decision-making in a fuzzy environment. Consider the following examples: x should be in the vicinity of x_0 where x_0 is a constant, and its analog: a particle is more or less spherical. The source of fuzziness in the mathematical statement is vicinity and in the verbal statement it is more or less.

The fundamental point here is that in much of science we assume that the source of imprecision is randomness. Clearly, we must guard against always making this assumption because in practice the source of imprecision may be fuzziness simply because our goals or our constraints or both are only loosely defined.

A. Properties of Fuzzy Sets[14]

There is a clear distinction between fuzziness and randomness. Thus, randomness is concerned with the degree of uncertainty of membership (or nonmembership) of a particular object in an ordinary set.* By contrast, fuzziness is concerned with classes in which there may be grades of membership intermediate between full membership and nonmembership. A fuzzy set, therefore, consists of a class of objects with a continuum of grades of membership. Such a set of characterized by a membership function which assigns to each object a grade of membership ranging from zero to one. The following specifics can be defined: an empty fuzzy set, the complement of a fuzzy set, containment, union of two fuzzy sets and intersection of two fuzzy sets. Additionally, many identities which obtain for ordinary sets can be extended to fuzzy sets: De Morgan's laws, Distributive laws. Finally many algebraic operations can be conducted on fuzzy sets, as they can be on ordinary sets: algebraic product, algebraic sum, absolute difference.

The symbols U, V, etc. are used to denote a specific universe of discourse, such as a set of persons who are living in a town; the set of all of the nuclear weapons that we have; the set of all real, even numbers, etc.

If A is a nonfuzzy subset of U and it consists of members u_1, u_2, . . . , then

$$A = u_1 + u_2 + \ldots + u_n \tag{47}$$

in which the plus sign denotes and.

If we let F be a finite fuzzy subset of U, then we can say that

$$F = \frac{\mu_1}{u_2} + \ldots + \frac{\mu_n}{u_n} \tag{48}$$

in which μ_i/μ_i is the grade of membership of u_1 in F, and $\mu_i = 1,2, \ldots, O = \leq\mu_i\leq 1$, $\mu_i = 0$ denotes no membership, and $\mu_i = 1$ denotes full membership. For example: if

$$u = a + b + c + d \tag{49}$$

* By the term ordinary set we mean a nonfuzzy set.

then we may have

$$A = a + b + d, \text{ and } F = 0.3a + 0.9b + d \quad (50)$$

And if

$$u = 0 + 0.1 + 0.2 \quad (51)$$

then, for example:

$$F = \frac{0.3}{0.5} + \frac{0.6}{0.7} + \frac{0.8}{0.9} + \frac{1}{1} \quad (52)$$

i.e., the grade of membership of 0.5 in u is 0.3 and so on.

Two types of membership functions have been distinguished:

$$S(u, \alpha, \beta, \gamma) \quad (53)$$

$$\pi(u, \beta, \gamma) \quad (54)$$

These are shown in Figure 45.

If F and G are fuzzy subsets of U, their union F ∪ G intersection F ∩ G bounded-sum F + G, and bounded-difference F ⊖ G, are fuzzy subsets of U defined by:

$$F \cup G \triangleq \int_U \mu_F(u) \vee \mu_G(u)/u \qquad \text{Union} \quad (55)$$

$$F \cap G \triangleq \int_U \mu_F(u) \wedge \mu_G(u)/u \qquad \text{Intersection} \quad (56)$$

$$F \oplus G \triangleq \int_U 1 \wedge (\mu_F(u) + \mu_G(u))/u \qquad \text{Bounded-sum} \quad (57)$$

$$F \ominus G \triangleq \int_U 0 \vee (\mu_F(u) - \mu_G(u))/u \qquad \text{Bounded-difference} \quad (58)$$

in which ∨ and ∧ denote max and min, respectively. Δ signifies ν is defined as; ν signifies the union of fuzzy singletons μ_i. For example with u defined as in Equation 49,

$$F = \frac{0.4}{a} + \frac{0.9}{b} + \frac{1}{d} \quad (59)$$

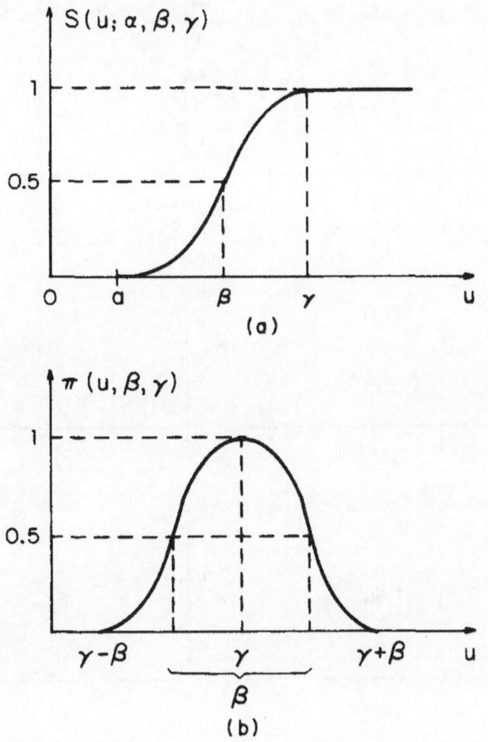

FIGURE 45. Plots of S and π functions. (From Zadeh, L., *Inf. Control,* 8, 338, 1965. With permission.)

$$G = \frac{0.6}{a} + \frac{0.5}{b} \tag{60}$$

then

$$F \cup G = \frac{0.6}{a} + \frac{0.9}{b} + \frac{1}{d} \tag{61}$$

$$F \cap G = \frac{0.4}{a} + \frac{0.5}{b} \tag{62}$$

$$F \oplus G = \frac{1}{a} + \frac{1}{b} + \frac{1}{d} \quad \text{and} \tag{63}$$

$$F \ominus G = \frac{0.4}{b} + \frac{1}{d} \tag{64}$$

Also, the compliment of F, designated F′, is defined by:

$$F' = \int_U \frac{1-\mu_F(u)}{u} \tag{65}$$

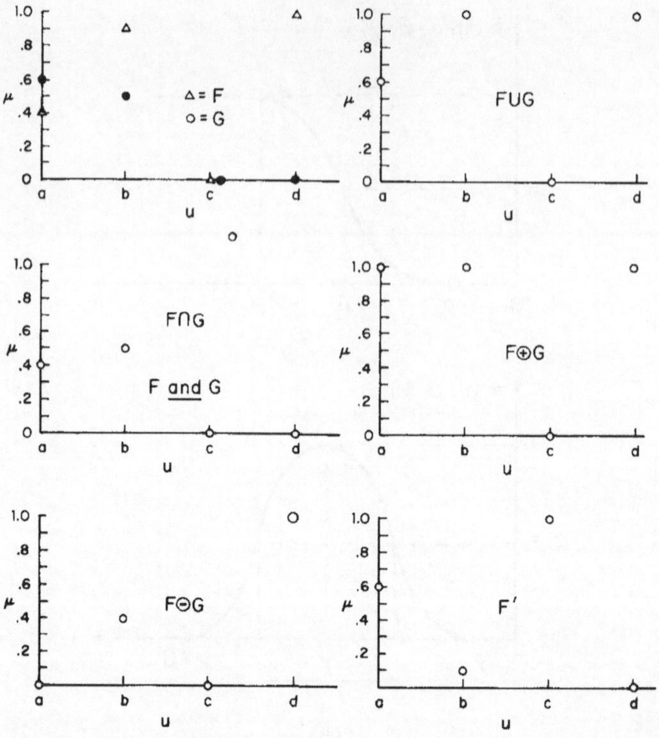

FIGURE 46. Graphical illustrations of operations on fuzzy sets. Refer to Equations 55 through 66.

From 59 therefore, F′ is in that case

$$F' = 0.6a + 0.1 + c \tag{66}$$

Equations 61 through 64 and 59 are plotted in Figure 46. If α is a real number, then F is defined as:

$$F^{\alpha} \triangleq \int \frac{(\mu_F(n))}{u} \tag{67}$$

In the case of fuzzy set F of 59

$$F^2 = \frac{0.16}{a} + \frac{0.81}{b} + \frac{1}{d} \tag{68}$$

$$F^{1/2} = \frac{0.2}{a} + \frac{0.3}{b} + \frac{1}{d} \tag{69}$$

F^2 and $F^{1/2}$ are illustrated in Figure 47.

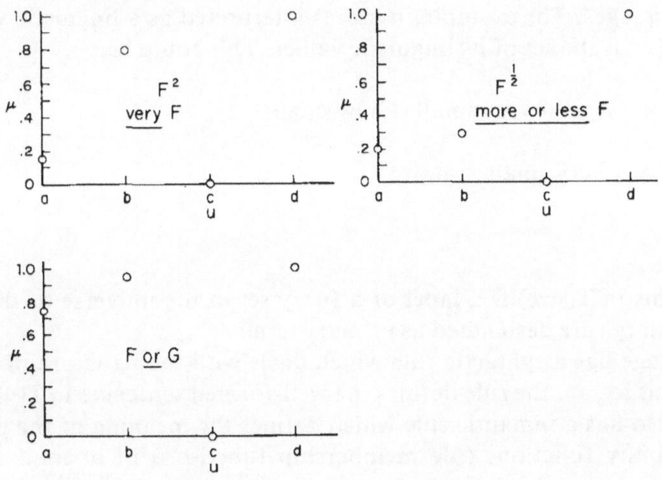

FIGURE 47. Graphical illustrations of operations on fuzzy sets. Refer to Equations 68 through 71.

B. Linguistic Labels

The operations which have just been demonstrated can be associated with appropriate linguistic labels.

For example, the operation of intersection of two fuzzy sets F, G may be identified with the label and:

$$F \text{ and } G \triangleq F \cap G \tag{70}$$

similarly

$$F \text{ or } G = F + G = \int_U \frac{\mu_F + \mu_G - \mu_F\mu_G}{u} \tag{71}$$

and

$$\text{very } F = F^2 \tag{72}$$

also,

$$\textit{more or less } F = F^{1/2} \tag{73}$$

F or G for the fuzzy sets defined in Equation 59 and Equation 60 is shown in Figure 47.

C. Linguistic Variables

A linguistic variable v is one whose values are words or sentences in natural or artificial verbal languages. For example, if size is interpreted as a linguistic variable, then its term's set, T(v), is the set of its linguistic values. This could be:

T(size) = small + large + very small + not small

+ very large + very very small + more

or less small + . . .

Each of the terms in T(size) is a label of a fuzzy set in the universe of discourse. The terms small and large are designated as primary terms.

Now, a language has a syntactic rule which deals with the structure of the language generated. That is to say, the rule defines the well-formed sentences in T(v).

A language also has a semantic rule which defines the meaning of the terms in T(v). If the compatability functions (the membership functions) of a small and large are designated by μ small and μ large, respectively, then from Equations 65, 72, and 73 we can derive the compatability functions of the nonprimary terms as follows:

$$\mu \textit{ very small} = (\mu \text{ small})^2 \tag{74}$$

$$\mu \textit{ more or less large} = (\mu \text{ large})^{1/2} \tag{75}$$

$$\mu \textit{ not very small} = 1 - (\mu \text{ small})^2 \tag{76}$$

Plots of the compatability (membership) functions of these terms are given in Figure 48[14] for purposes of illustration. In this connection, it should be pointed out that the problem of measuring the fuzzy set is not dealt with to any major extent in the literature. It has been observed, however, that "Probably, we should not expect particular numerical values of shortness to be meaningful (except 0 and 1), but rather that their ordering. Observers may agree on whether one man is shorter than another. But their use of the word short will never stabilize".[41] It is now easier to understand the three attributes of fuzzy sets — generality, ambiguity, and vagueness.

D. Geometric Comparator

A geometrical comparator of simple design has been constructed as shown in Figure 49. The principle behind this assembly is that the various shapes are developed by means of simple operations upon equidimensioned shapes. For example with a starter equidimensioned pentagon, the shape is extended to different values of ratio height to base (2,4,8 and 10 to 1, respectively). Similarly, the pentagon starter is contracted in ratios of height to base of ½, ¼, 1/8, and 1/10, respectively. The third modification is carried out by a skew at 30° and 60°, respectively. The chart shown in Figure 49 is developed after duplications are eliminated. This method is a convenient way of developing a rational variety of geometric profiles. The comparator has been used with modest success in the identification of examples from the Particle Atlas and also in the identification of a published set of limestone chips. However, the agreement as to assignment of category often differs for different observers. This is presumably because each observer calls upon a different set of mental images (templates) due to their

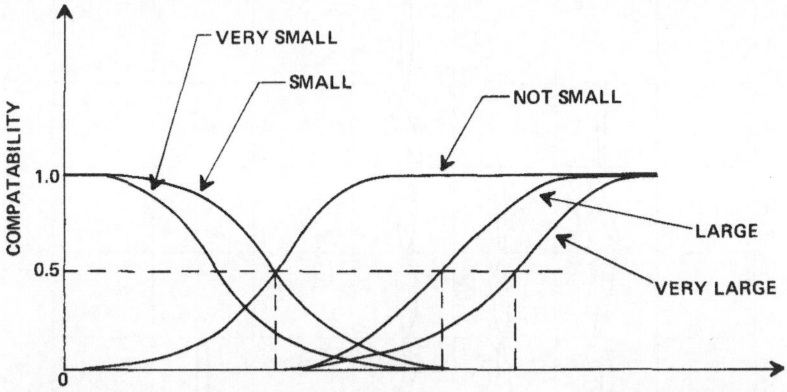

FIGURE 48. Compatability (membership) functions for small and large. (From Zadeh, L., Fu, L., Tanaka, K., and Shimura, M., *Fuzzy Sets and their Applications to Cognitive and Decisiion Processes,* Academic Press, New York, 1975. With permission.)

having had different living experiences. The problem therefore resolves to one of careful definition of the variation of geometric shape in relationship to the way in which we express shape. But how does one carry out this operation? Three approaches are discussed below.

The principle underlying this operation may be illustrated as follows. Let x be a space of objects with generic element x. Thus $X = \{x\}$. A fuzzy set A in X is characterized by a membership function $f_a(x)$ which associates with each object in X a real number in the interval [0,1], with the value of $f_A(x)$ at x representing the grade of membership of x in A.

The intersection of two fuzzy sets A and B with membership functions $f_A(x)$ and $f_B(x)$, respectively, is a fuzzy set $C = A \cap B$.

C has a membership function:

$$f_C(x) = \text{Min}\,[f_A(x), f_B(x)] \qquad x \in X \tag{77}$$

$$f_C = f_A \wedge f_B \tag{78}$$

A graphical representation of the intersection of two fuzzy sets in X is illustrated in Figure 50. The heavy lines denote the intersection. Note that x belongs to fuzzy set A if $f_A(x)$ is positive but that its degree of membership is variable relative to others in the set. (Compare this to an ordinary set in which either x belongs here $f_A(x) = 1$; or x does not belong here $f_A(x) = 0$). More importantly, in the case of a fuzzy set if $f_A(x)$ is positive, what we really want to know is its grade of membership. In this sense we need to know not just that it belongs but to what extent it does so.

An example of the intersection of two fuzzy sets comprising cube and sphere is represented in Figure 51. Note that this is a two-dimensional representation. The grades of membership depicted are quite arbitrary and within the confines of the example, effectively constitute scale divisions.

Another interesting approach due to Lee considers the classification of triangles and quadrangles.[42] The attempt with triangles is quite successful but this is not so with quadrangles. The latter have too many degrees of freedom for this treatment. Five

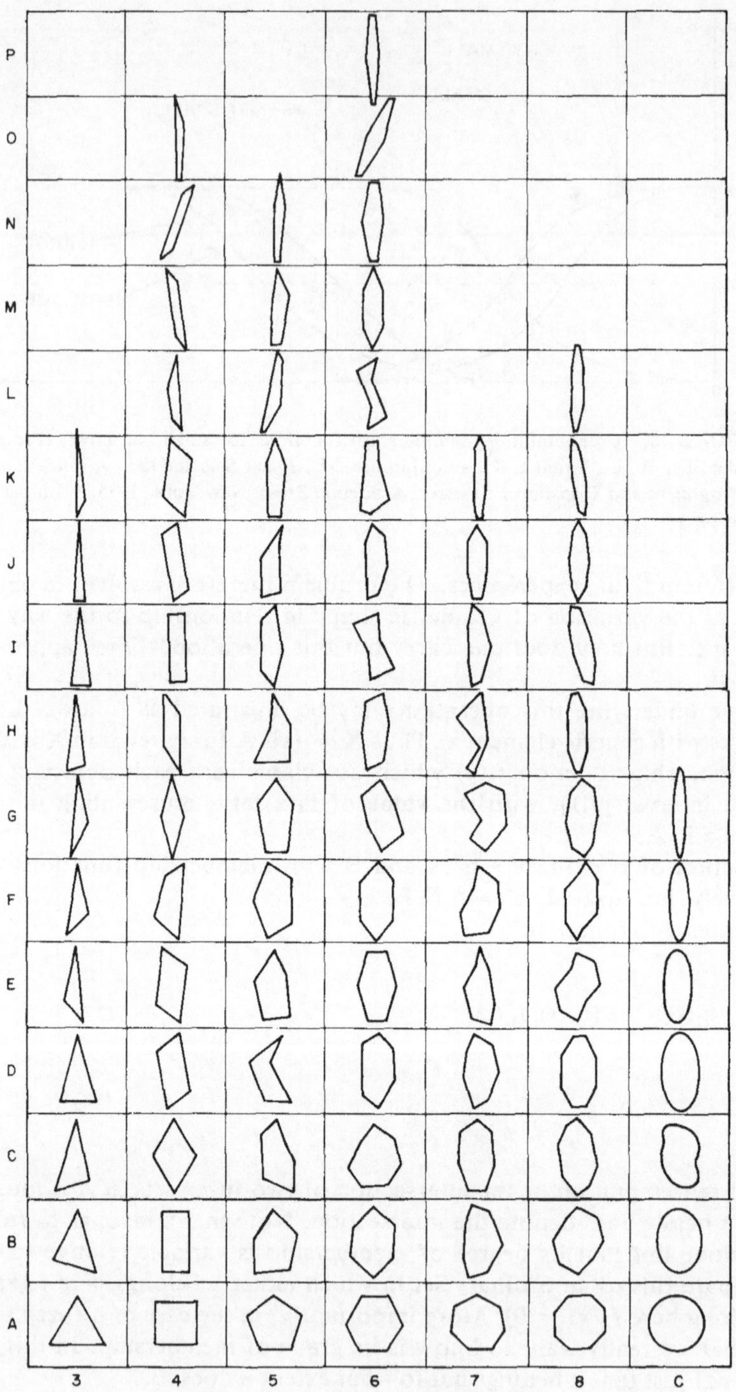

FIGURE 49. Geometrical comparator.

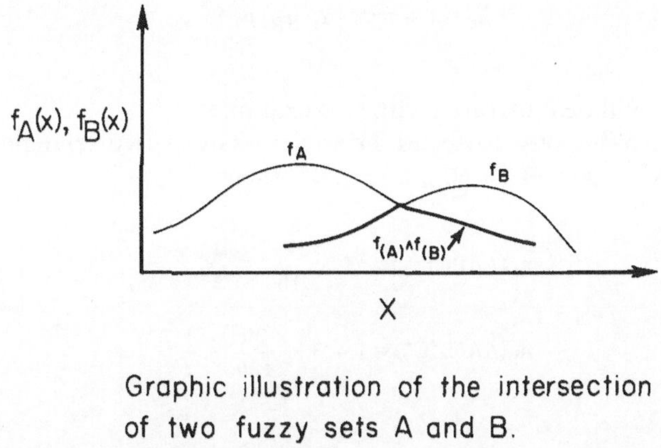

FIGURE 50. Graphic illustration of the intersection of two fuzzy sets A and B. (From Beddow, J. K., *J. Res. Assoc. Powder Technol. Jpn.*, 15(4), 227, 1978. With permission.)

different triangles are defined: approximate isosceles triangles, approximate equilateral triangles, approximate right triangles, approximate isosceles right triangles, and ordinary triangles (i.e., the rest of them). These terms are defined in the following way: given a triangle ABC with angles A, B, and C. An approximate isosceles triangle may be expressed by a grade of membership:

$$\mu_1 (ABC) = 1 - \rho_1 \min \{ |A - B|, |B - C|, (C - A) \} \quad (79)$$

in which ϱ_1 is a constant and is 1/60.
An approximate equilateral triangle may be expressed by a grade of membership:

$$\mu_E = 1 - \rho_E \max \{ |A - B|, |B - C|, |C - A| \} \quad (80)$$

in which ϱ_E is a constant and is 1/180.
An approximate right triangle may be expressed by a grade of membership:

$$\mu_R = 1 - \rho_R \min \{ |A - 90^\circ|, |B - 90^\circ|, |C - 90^\circ| \} \quad (81)$$

in which ϱ_R is constant and is 1/90.
An approximate isosceles right triangle may be expressed by a grade of membership:

$$\mu_{IR} = 1 - \max \left\{ \frac{\min [|A - B|, |B - C|]}{60^\circ}, \frac{|A - 90^\circ|}{90^\circ} \right\} \quad (82)$$

An ordinary triangle may have a grade of membership defined as

$$\mu_0 = 1 - \max\,[\mu_I, \mu_R, \mu_E]$$

These relationships will be illustrated with two examples.

Example 1: Approximately Isosceles Triangles—Given two triangles: A = 90°, B = 70°, C = 20°, A′ = 90°, B′ = 60°, C′ = 30°, then

$$\mu_I\,(\Delta ABC) = 1 - 3 \times \frac{20}{180} = \frac{2}{3}$$

$$\mu_I\,(\Delta A'B'C') = 1 - 3 \times \frac{30}{180} = \frac{1}{2}$$

This result is intuitively appealing, as shown in Figure 52.

Example 2: Approximate Right Triangles—Given two triangles: A = 45°, B = 75°, C = 60°, A′ = 60°, B′ = 60°, C′ = 60°, then

$$\mu_R(\Delta ABC) = \frac{5}{6}$$

$$\mu_R(\Delta A'B'C') = \frac{2}{3}$$

The graph representation of ΔABC and ΔA′B′C′ is shown in Figure 53.

The problem with trying to classify four-sided figures using this approach seems to be that both angles and side lengths have to be taken into account so that equivocal results are obtained. As the number of sides increases this requirement may tend to become something of a burden.

A third approach that is yet to be studied is to use the values of the coefficients from the Fourier (or Walsh) analysis in order to define a degree of membership in a particular class. For example, triangularity may be represented by the value of A_3. This procedure requires some rules for normalization so that the degree of membership of unity can be a benchmark.

E. Verbal Definitions of Complex (Fuzzy) Concepts[43]

A complex concept usually requires a definition. This definition will vary depending on its origin. Conventional approaches include: a dictionary definition, writing an essay, approximating to the concept by a less fuzzy concept and giving a precise definition for the latter. A newly proposed approach involves the development of a fuzzy algorithm in the form of a branching questionnaire. In the example shown in Figure 54, the composite question Q Δ B consists of questions Q_1, Q_2, . . . ,Q_n which are asked in order. This order is determined by the answers to the previous questions. The tableau in Figure 55 is a more conventional block diagram representation.

As an illustrative example of an admittedly complex, fuzzy phenomenon, the linguistic relational representation for recession has been shown for a two-quarter period. This is given in Table 17.[43]

There are many nuances in this particular approach that have to be carefully studied but the method appears to be applicable to fine particle characterization.[44] In addition, it may be advantageous to utilize this approach in connection with the relationships between particle shape and mode of origin and particle shape and behavior, respectively.

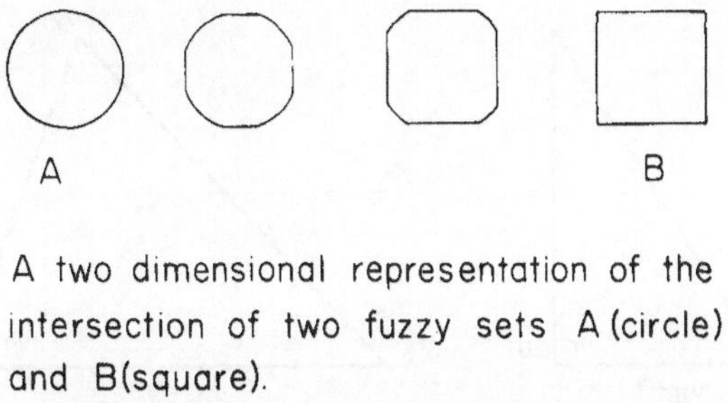

FIGURE 51. A two dimensional representation of the intersection of two fuzzy sets A (circle) and B (square). (From Beddow, J. K., *J. Res. Assoc. Powder Technol. Jpn.*, 15(4), 227, 1978. With permission.)

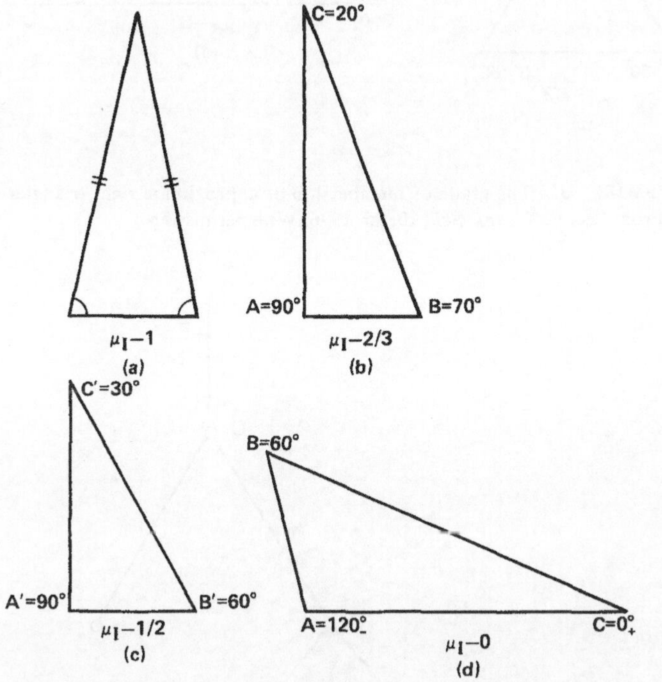

FIGURE 52. The grade of membership of approximate isoceles triangles. The minus sign beside a number indicates a quantity slightly less than shown, a plus sign indicates a quantity slightly greater than shown. (From Lee, E. T., Inf. Sci., 10, 95, 1976. With permission.)

VIII. SHAPE-TEXTURE ANALYSIS OF FINE PARTICLES

Shape deals with the external pattern of the particle whereas texture pertains to the internal pattern of the particle. One can state the following hypothesis: *There exists a set of relationships between the morphology of a particle and the texture. The texture describes the internal chemical and structural detail of the particle and the morphology describes the size-shape characteristics of the particle.*

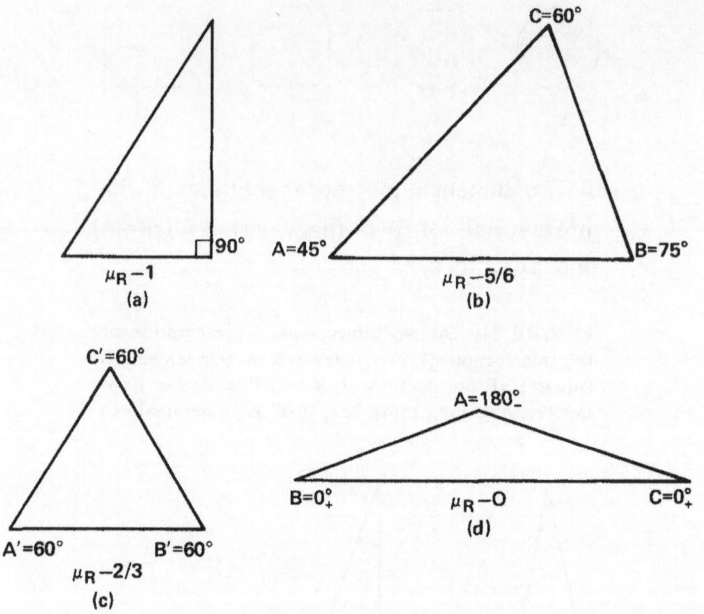

FIGURE 53. The grade of membership of approximate right triangles. (From Lee, E. T., *Inf. Sci.*, 10, 95, 1976. With permission.)

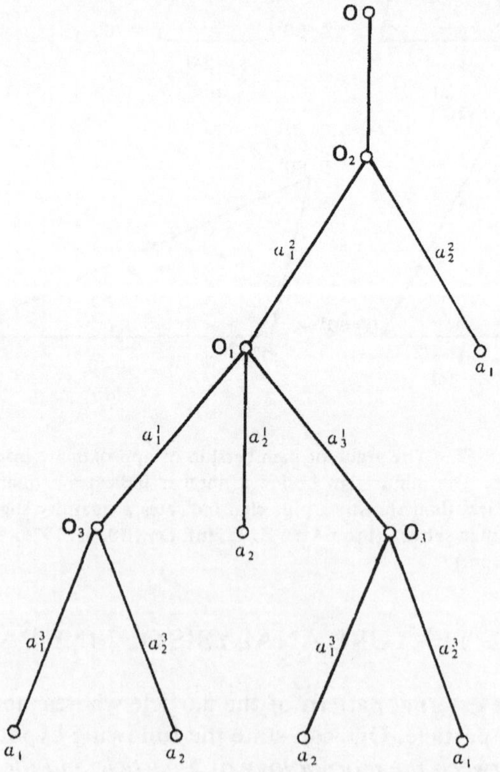

FIGURE 54. An example of branching questionaire. (From Zadeh, L. A., *Int. Man-Mach. Stud.*, 8, 249, 1976. With permission.)

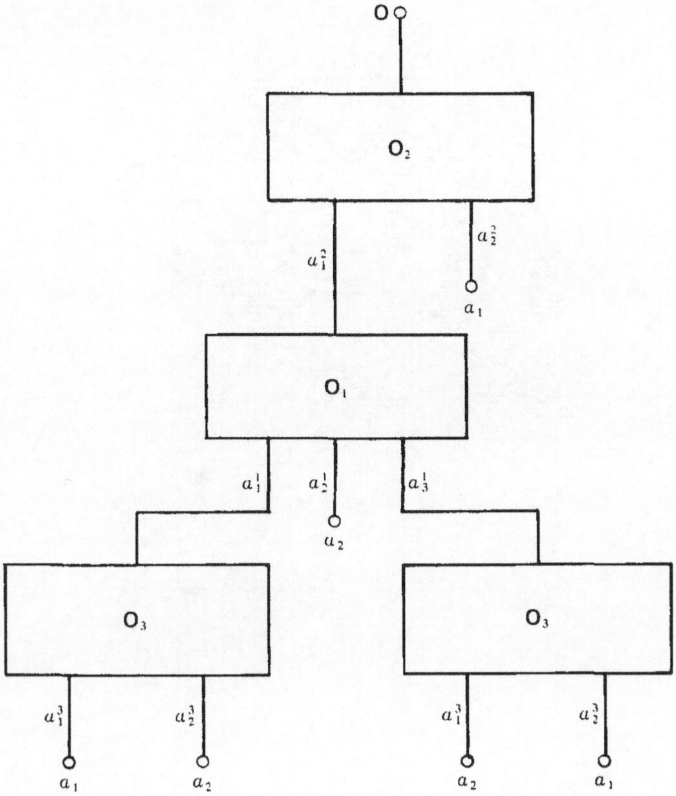

FIGURE 55. Block diagram representation of a branching questionaire. (From Zadeh, L. A., Int. Man-Mach. Stud., 8, 249, 1976. With permission.)

TABLE 17

Tableau of Relational Representation of Recession

Decline in gross national product	Unemployment	Increase in bankruptcies	Decline in Dow Jones Index[a]	Recession
Small	Low	Small	Small	False
Moderate	Low	Small	Small	Not true
High	Low	Small	Small	Borderline
High	Moderate	Moderate	Large	Rather True
High	High	Large	Large	Very True

[a] Relative to its maximum for the interval.

From Zadah, L. A., Int. J. *Man-Machine Studies,* 8, 249, 1976. With permission.

In biological systems it is possible to discern internal structures with microscopic methods currently available. In other systems — such as minerals, metals, pigments, and the like, additional techniques are necessary. As an illustrative point, recent work in multicompositional particle characterization using the semimicroprobe in order to determine the sizes and compositions of complex particles has been reported. Figure

A

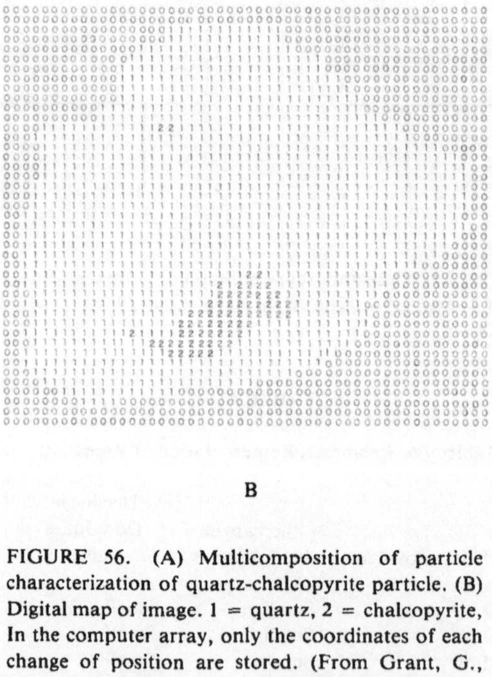

B

FIGURE 56. (A) Multicomposition of particle characterization of quartz-chalcopyrite particle. (B) Digital map of image. 1 = quartz, 2 = chalcopyrite, In the computer array, only the coordinates of each change of position are stored. (From Grant, G., Hall, J. S., Reid, A. F., and Zuiderwyk, M., *Scanning Electron Microscopy/1976,* (Part III), IIT Research Institute, Chicago, 1977, 401. With permission.)

56[45] shows some results of an analysis of a particle of quartz-chalcopyrite composition. Figure 56A shows the raster image obtained with back scatter electrons and the bottom image is a digital map of the image. The regions of quartz and chalcopyrite are clearly shown. The computed particle parameters are given in the accompanying data table (Figure 56B). It would not be impossible to interrelate this chemical analysis with morphological analysis and this would represent a significant move in the direction of a complete characterization of a fine particle.

IX. RECENT DEVELOPMENTS

A. Morphology and Mode of Particle Origin

It has been hypothesized that particles with similar genesis will possess similar morphologies.[5] For example, they will have similar signatures. This hypothesis has not been tested. We have prepared a set of particles of adipic acid by the following methods: atomization using air, comminution, and crystallization. Sample microphotographs are shown in Figure 57. At the present time these images are being digitized.

B. Morphology and Chemical Reaction

It is well known that particle morphology influences chemical interactions between the particle and its environment. We are at an early stage in the development of techniques to study these relationships. In the first instance we have used fairly large particles in order to simplify handling problems. Later in the program we plan to use smaller particles for study.

In essence, the technique consists of passing a particle through a reaction zone and quenching the reaction as it passes from the zone. The profile is digitized before and after and the results subject to analysis. In the first experiment, the reaction zone is a layer of acid through which the particle moves under gravity. It is quenched in alkaline solution when it passes down and out from the acid layer. We developed the (Φ,ℓ) program (Section IIIB) in order to be able to analyze reentrant particles.

An interesting sequence of pictures which shows the effects of chemical attack upon particle morphology is shown in Figures 58 through 62, respectively. It is very interesting to note that the function of the sum of the coefficients (in this case Fourier coefficients are being used) clearly demonstrates in Figure 63 primary stage of chemical attack (corresponding to a smoothing of the particle outline) and also the secondary attack (corresponding to the intrusion of the chemical reagent (the HCl)) into the body of the zinc particle. It is also interesting to note that the more traditional shape factor approach (such as perimeter divided by area) does not pick up this secondary attack effect.[27]

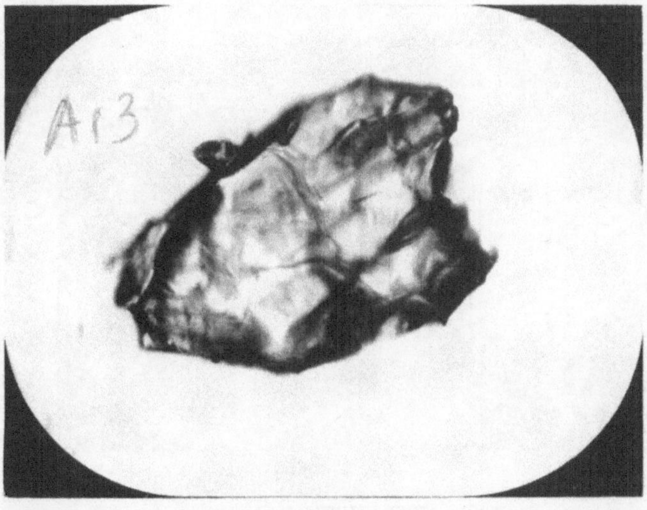

A

FIGURE 57. Adipic acid produced by (A) atomizing, (B) comminution, (C and D) crystillization.

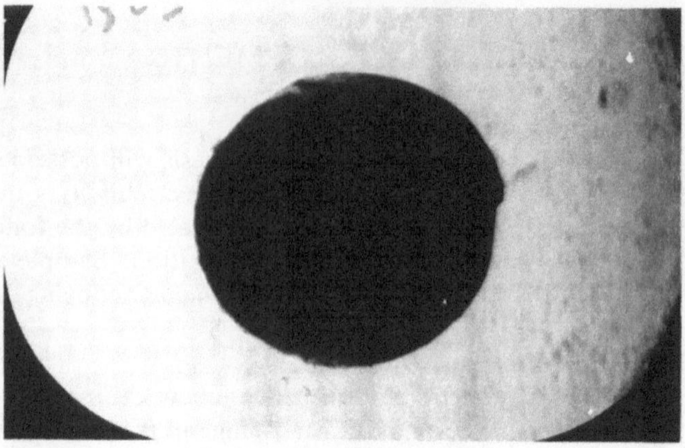

FIGURE 57B

FIGURE 57C

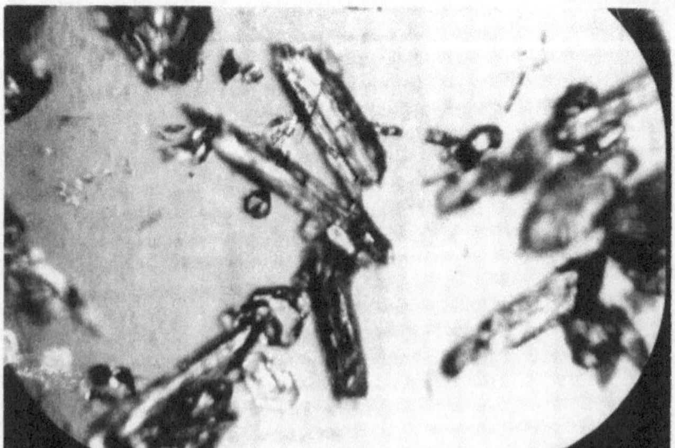

FIGURE 57D

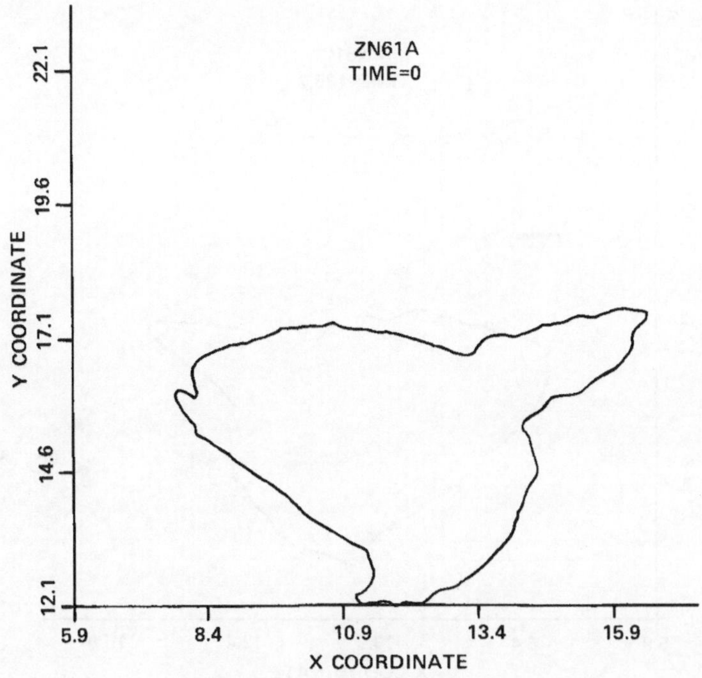

FIGURE 58. Original shape of Zn61.

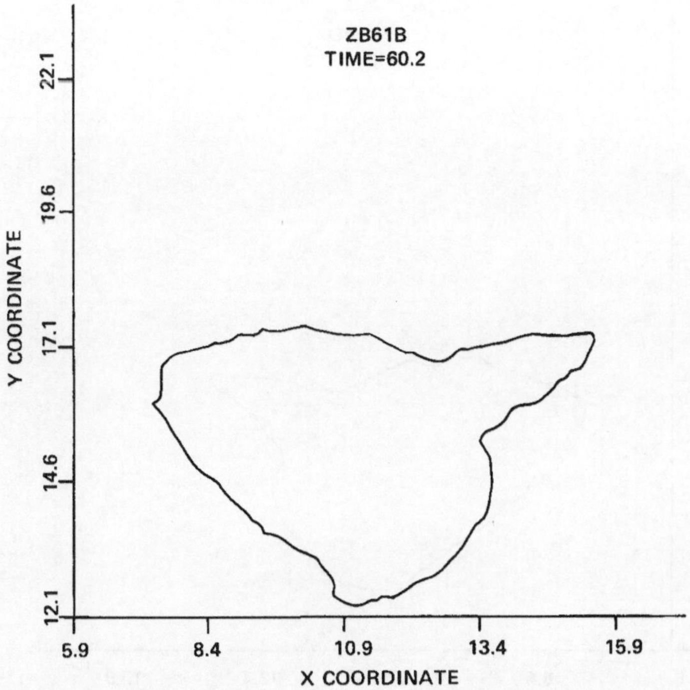

FIGURE 59. Shape of Zn61 after 60.2 sec.

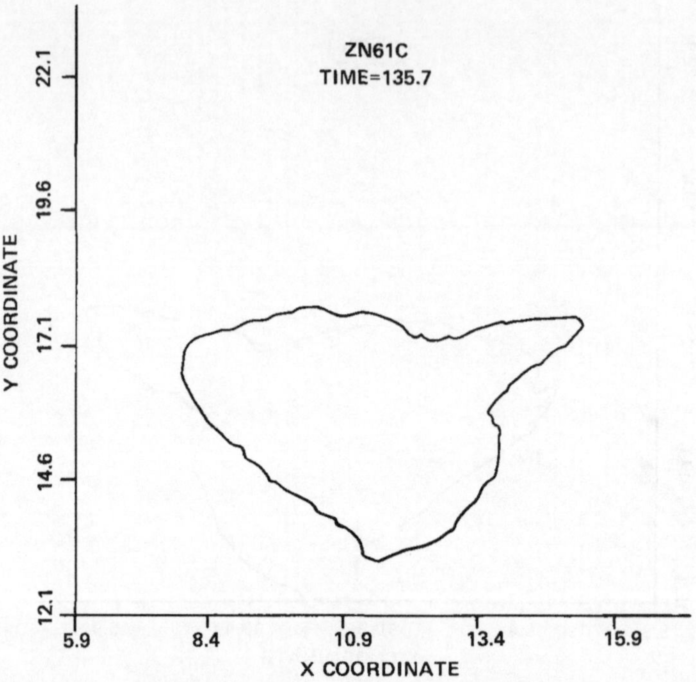

FIGURE 60. Shape of Zn61 after 135.7 sec.

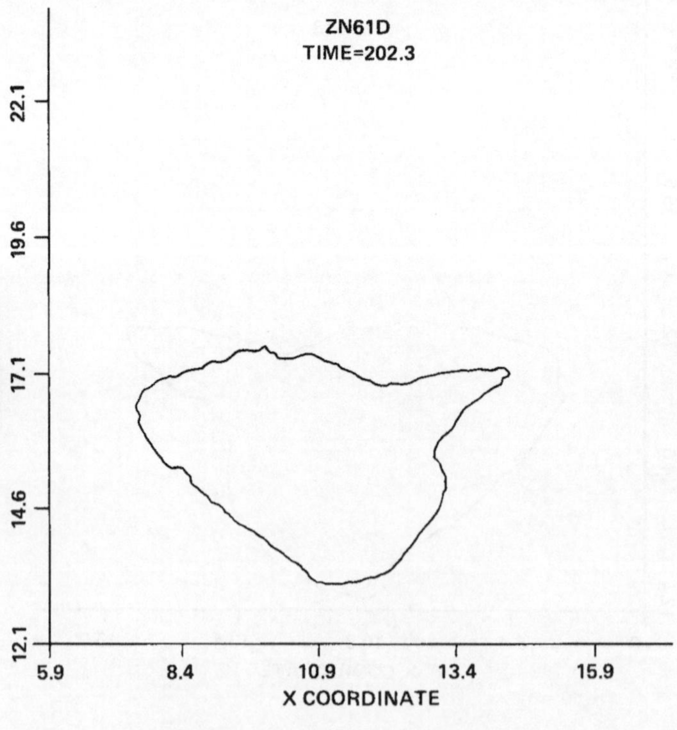

FIGURE 61. Shape of Zn61 after 202.3 sec.

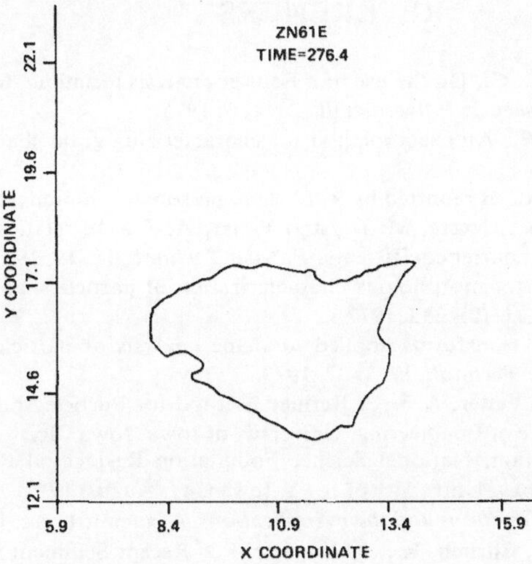

FIGURE 62. Shape of Zn61 after 276.4 sec.

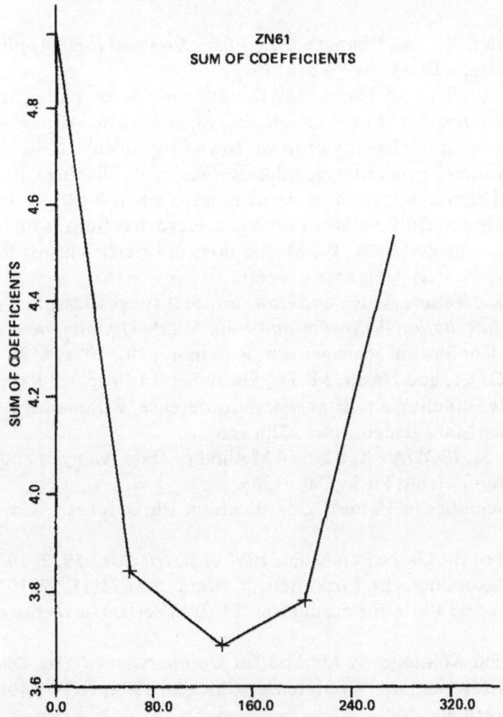

FIGURE 63. Plot of sum of coefficients time for Zn61. (From Fong, S-T., Beddow, J. K., and Vetter, A. F., Powder Technol., submitted, 1979. With permissiom.)

REFERENCES

1. **Beddow, J. K. and Philip, G. C.**, On the use of a Fourier analysis technique for describing the shape of individual particles, *Planseeber. Pulvermetall.*, 23, 1, 3, 1975.
2. **Ehrlich, R. and Weinberg, B.**, An exact method for characterizing grain shape, *J. Sed. Petrol.*, 40, 1, 205, 1970.
3. **Redmond, B. and Walz, S. R.**, as reported by R. Ehrlich, personal communication, May, 1978.
4. **Beddow, J. K., Philip, G. C., Nasta, M. D., and Vetter, A. F.**, On relating some particle profile characteristics to the profile Fourier coefficients, *Powder Technol.*, 18, 19, 1977.

5a. **Meloy, T. P.**, A hypothesis for morphology characterization of particle shape and physicochemical properties, *Powder Technol.*, 16(2), 233, 1977.

5b. **Meloy, T. P.**, Fast Fourier transforms applied to shape analysis of particle silhouettes to obtain morphological data, *Powder Technol.*, 17, 1, 27, 1977.

6. **Fong, S-T., Beddow, J. K., Vetter, A.F.**, A Refined Method for Particle Shape Representaton, Report No. G 178 78 02, College of Engineering, University of Iowa, Iowa City.
7. **Redinbo, R.**, oral presentation, National Science Foundation Residential Research Workshop on Advanced Particle Morphology, University of Iowa, Iowa City, August 1977.
8. **Beauchamp, K. G.**, *Walsh Functions and their Applications*, Academic Press, London, 1975.
9. **Yarus, J. M., Brown, P. J., Ehrlich, R.**, A New Tool For Recent Sediment Studies, Fourier Grain Shape Analysis, American Geological Union, Spring Meeting 1978, Report number 029248 Ehrlich.
10. **Pryzgock, R. S.**, Identification of Sediment Sources in a Fluvial Network — North Carolina Blue Ridge — Fourier Grain Shape Analysis, Master's thesis, University of South Carolina, 1976.
11. **Meloy, T. P.**, personal communication, 1978.
12. **Davies, R.**, Feature-space representation, *Powder Technol.*, 12(12), 111, 1975.
13. **Tou, J. T. and Gonzalez, R. C.**, *Pattern Recognition Principles*, Addison-Wesley, Reading, Mass., 1974.
14. **Zadeh, L., Fu, K., Tanaka, K., and Shimura, M.**, *Fuzzy Sets and their Application to Cognitive and Decision Processes*, Academic Press, New York, 1975.
15. **Beddow, J. K., Philip, G. C., and Nasta, M. D.**, On the use of pattern recognition methods in particle shape analysis, *Planseeber. Pulvermetall.*, 24, (3), 17, 1976. Also in Report No. G190-NSF - 76-001, College of Engineering, University of Iowa, Iowa City, Iowa.
16. **Underwood, E. E.**, *Quantitative Stereology*, Addison-Wesley Publishing Co., Reading, Mass., 1970.
17. **Beddow, J. K.**, Physical characterization of particulates, *Powtech '77, Proceedings of 4th International Powder Technology and Bulk Solids Conference*, Heyden & Sons, London, 1977, 99.

17a. **Lew, G., Beddow, J. K., and Vetter, A. F.**, Morphological Coefficients of Particle Profiles Derived from Walsh Coefficients, *Powder Met. Int.*, in press.

18. **Wang, C., Beddow, J. K., Vetter, A. F., and Lew, G.**, Statistical Classifiers in Particle Morphology Analysis, *Proceedings: International Powder and Bulk Solids Handling and Processing Conference*, Industrial and Scientific Conference Management, Rosemont, Ill., 1978, 158.
19. **Beddow, J. K., Philip, G. C., and Nasta, M. D.**, Fuzzy Sets in Property Representation, Internation Powder and Bulks Solids Handling and Processing Conference, Rosemont, Ill., 1977, 391, Industrial And Scientific Conference Management, Inc., Chicago.
20. **Gall, G. H. and Hall, D. J.**, ISODATA, a Novel Method of Data Analysis and Pattern Classification, Stanford Research Institute, Menlo Park, Cal., 1965.
21. **Bezdek, J.**, Fuzzy Mathematics in Pattern Classification, Ph.D. Thesis, Cornell University, Ithaca, N.Y., 1973.
22. **Anderson, E.**, The Irises of the Gasce Peninsula, *Bull. Am. Iris Soc.*, 59, 2, 1935.
23. **Bezdek, J.**, Numerical Taxonomy with Fuzzy Sets, *J. Math. Biol.*, 1(1), 57, 1973.
24. **Chang, C. V.**, Fuzzy Sets and Pattern Recognition, Ph.D. Thesis, University of California, Berkeley, 1967.
25. **Goldman, A.**, A Modified Kolmogorov Method for Comparison of Size Distribution, Presented at Powder and Bulk Solids International Conference, Chicago, 1978, to be published by Industrial Scientific Conference Management, Inc.
26. **Anderson, T. W.**, *An Introduction to Multivariate Statistical Analysis*, John Wiley & Sons, Inc., 1958, 126.
27. **Fong, S-T., Beddow, J. K., and Vetter, A. F.**, Some Effects of Chemical Reaction Upon Particle Morphology, in preparation.
28. **Fix, E. and Hodges, J. L., Jr.**, Discriminant Analysis, Nonparametric Discrimination: Consistency Properties, U.S. Air Force School of Aviation Medicine, Project No. 21-49-004, Report No. 4, 1951.
29. **Fu, K. S., and Chou, C. K.**, Crop classification research, *Report on the 1968 IEEE Workshop on Pattern Recognition*, Chandrasekaran, B., Kanal, L., and Nagy, G., Eds., Institute of Electricians and Electronic Engineers, New York, 1969.

30. **Hall, E. L., Preston, K., and Roberts, F. E.**, Classification of Profusion of Black Lung Disease form Chest X-Rays, Proceedings First International Joint Conference on Pattern Recognition, 1973, Washington, D. C.,
31. **Gilbert, E. S.**, On Discrimination Using Qualitative Variables, *JASA*, 63, 1399, 1968.
32. **Lachenbruch, P. A., Sneering, D., and Revo, L. T.**, Robustness of Linear and Quadratic Discriminant Functions to Certain Types of Non-Normality, *Commun. Stat.*, 1, 57,1973.
33. **Broffit, J. D., Randles, R. H., and Hogg, R. V.**, Distribution-free Partial Discriminant Analysis, *JASA*, 71, 934, 1976.
34. **Lachenbruch, P. A., and Mickey, M. R.**, Estimation of Error Rate in Discriminant Analysis, *Technometrics*, 10, 1, 1968.
35. **Adey, W.**, Organization of Brain Tissue: Is the Brain a Noisy Processor? *Int. J. Neurosci.*, 3, 271, 1972.
36. **Beddow, J. K.**, On the Design of a System for Particle Shape Analysis, *J. Res. Assoc. Powder Technol., Jpn.*, 15(4), 227, 1978.
37. **Rao, P.**, Shape and Other Properties of Gas Atomized Metal Powders, Ph.D. Thesis, Drexel University, 1973.
38. **Lowrison, G. C.**, *Crushing And Grinding*, CRC Press, Cleveland, 1974.
39. **Zadeh, L.**, Fuzzy Sets, *Inf. Control.*, 8, 338, 1965.
40. **Bellman, R. E. and Zadeh, L.**, Decision Making in a Fuzzy Environment, *Manage. Sci.*, 17(4), B141, 1970.
41. **Goguin, J. A.**, The Logic of Inexact Concepts, *Systhese*, 19, 325, 1968-69.
42. **Lee, E. T.**, An Application of Fuzzy Sets to the Classification of Geometric Figures and Chromosome Images, *Inf. Sci.*, 10, 95, 1976.
43. **Zadeh, L. A.**, A Fuzzy Algorithmic Approach to the Definition of Complex or Imprecise Concepts, *Int. J. Man-Mach. Stud.*, 8, 249, 1976.
44. **Zadeh, L. A.**, personal communication.
45. **Grant, G., Hall, J. S., Reid, A. F., and Zuiderwyk, M.**, Multicompositional particle characterization using the SEM-microprobe, *Scanning Electron Microscopy/1976*, (Part III), IIT Research Institute, Chicago, 1976, 401.
46. **Beddow, J. K., and Vetter, A. F.**, On the use of classifiers in morphological analysis of particulates, *J. Powder and Bulk Technol.*, 1(1), 42, 1977.
47. **Lew, G., Beddow, J. K., and Vetter, A. F.**, Morphological coefficients of particle profiles derived from Walsh coefficients, submitted to *Powder Metallurgy Int.*, in press, 1979.
48. **Fong, S-T., Beddow, J. K., and Vetter, A. F.**, Acid zinc reaction predictably affects particle morphology, *Powder Technol.*, 23 (No.2), 187, 1979.

Chapter 2

SEARCH FOR SIGNATURES

T. P. Meloy

TABLE OF CONTENTS

I. INTRODUCTION

This section describes a new way of quantitatively characterizing particles. By particles I mean sand grains, soils, gravel, metalliferous ores, coal, cement, corn, wheat, salt, sugar, and tens of thousands of other materials. In an industrial as well as an agricultural society, particles are ubiquitous. Most of the feed stock to industry is in partiulate form. Most of our building materials — bricks, cement, metals — are or were at one time in particulate form. Of course an agricultural society depends on the fertility of the soils which are combinations of particles. The packing density of particles, the bearing strength, the angle of repose, and many other physical properties of particles are of importance to the engineer. Until recently these properties could not be predicted from measurements of particles.

One often forgets the vast extent of particulate material utilized by society. For example, 100 million tons of steel produced a year implies something like 400,000,000 tons of ore was dug, crushed, processed, and an equal amount of overburden removed. Our national goal of 1 to 1½ billion tons of coal per year means that in waste products there will be between 300 and 500 million tons of toxic waste to be disposed of. Copper production of 1 million tons a year means that 200 million tons of ore has been dug and the waste disposed of. If one considers construction and farming as working with particulates, then the total quantity of material handled by earth-moving equipment or plows is very large. One must never lose sight of the smaller industries like fertilizer, cement, and the ceramic industry to realize how pervasive and ubiquitous particles are. If the U.S. should be pushed into utilizing oil from oil shales or uranium from the Tennessee shales, then the waste materials alone will amount to cubic miles per year. While this is an area of great national importance, it is also an area that is not well understood.

A new method of quantitatively characterizing particles was developed independently by four investigators in the late 1960s. Schwartz,[1] Meloy,[2] Ehrlich,[3] and Beddow and Philip[4] independently applied Fourier series to describe particle shape. Using this technique, the particle shape can be broken down into its series of numbers, each of which has physical significance. This monograph describes the analytical techniques currently being used, some of the problems to which the technique has been applied, and the physiochemical significance of the resulting analysis. Interestingly enough, the approach for analyzing particles has raised deep, theoretical problems in mathematics. Furthermore, the technique proves to have applicability to a number of independent engineering fields. In addition, unexpected results are commonplace. For example, in taking a homogenous material such as quartz and crushing it, the result is not in one or two different particle shapes, but five different types of shapes.

Before describing the modern method of characterizing particles, it is best first to review the historical approaches to particle characterization. Most historical approaches were based on a sphere of equivalent diameter to the measured particle i.e., a sphere which would pass through a given size of screen or fall through a fluid at the same rate as the particle being measured. In other words, the particle size was equivalent to a sphere of the same density that would interact with the measuring instrument the same way. More recently, Quantamet® and other companies developed optical measuring techniques where the particle surface is scanned and yields the projected area of the particle and its perimeter. These techniques give one, or at best, two independent quantitative measures of the particle shape, one parameter of which invariably involves size. Other approaches used by Hausner[5] and Pah let al.[6] were used to give a series of standard particle shapes and compare the particle in question with one of the standard shapes. McCrone's atlas is this approach carried to an extreme. In the atlas, hundreds of different particles have been photographed and one need only take one's sample and compare it to one of the photographs to get an idea of the origin of the particle. McCrone's atlas is used primarily to identify airborne particles. Another approach is that of Davies,[7] who classified particles not only by size, but by blockiness, platyness, needleness, and roundness.

Ideally, any particle characterization technique should, at the very minimum, be capable of reproducing the particle shape beyond any specified error. In terms of industrial specifications or research, a series of numbers specifying the particle shape is a must. In many applications of particle characterization, particularly those involving specification, only a few of the characterizing numbers will be needed to define the particle. However, any useful characterization system must be able to reproduce the particle shape.

The question arises as to why does one have to describe particles by a series of numbers? Why can't one just say the particle is blocky, or chunky, or smooth like a robins egg, or some other descriptive comparison? Perhaps the best answer is that a blocky particle means different things to different people. Perhaps it is best to think of ordering a suit from a top London tailor and sending one or two dimensions. Suppose one sends one's height and weight. Would the suit fit very well? If one has ever been measured for a suit, one is astonished by the number of accurate mesurements that are taken. Furthermore, there are some 24 different posture shapes on the tailor's measurement form to take the place of unquantifiable measurements. Thus to get a decent tailor suit, a number of measurements must be made, and by the same token, to get a decent characterization of any individual particle, a number of optimal measurements must also be taken. In the field of modern technology, particulates is a very important field, but it had been held back because until recently no system of quantifying particle shapes had been developed.

It is sometimes useful to compare fluids and fluid systems with particulate systems. With fluids usually a few characteristics such as the pressure and temperature are sufficient to tell a great deal about how the system will behave. For example, with gases, molecular weight and shape have very little to do with gas behavior. With particles however, the problem is confounded because particle size and particle shape play an important role in the behavior of the system. As a result, the science of particulates should advance more rapidly with the application of the system described herein.

II. METHODS OF CHARACTERIZATION

To characterize a particle, one takes the projected area of the particle and records points on the particle outline. Think of a silhouette of a particle with the maximum projected area, taking discrete points along that silhouette edge. In Figure 1 the methodology is described. Using polar coordinates, one takes the radius versus that obtained in Figure 1 and uses discrete Fourier transforms to obtain the Fourier coefficients. These coefficients and their associated phase angles describe the particle silhouette beyond any level of error specified. Each of the coefficients has its own physical significance. For example, the zero coefficient is the particle radius; the first coefficient is the error in locating the center of area of the particle; the second coefficient is the measure of aspect ratio; and the third coefficient is the measure of triangularity of the particle. See Figures 2,3,4, and 5. Each of the higher-order coefficients have their own meaning; however, unless a coefficient has a relatively large amplitude, individual coefficients do not appreciably alter the particle shape. It is common to speak of the first 8 or 16 coefficients as containing the macroinformation, the 17th through the 128th as the mesocoefficient, and the high-order coefficient as being the

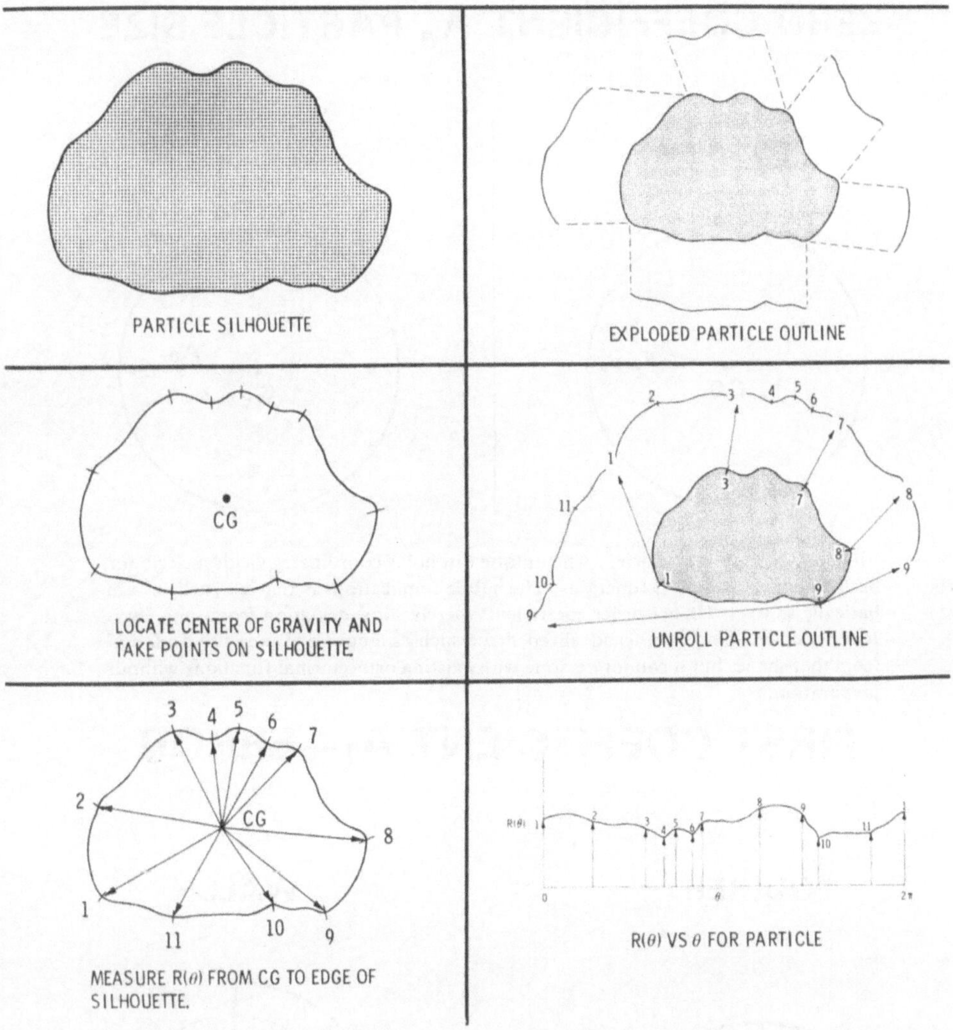

FIGURE 1. Transformation of particle outline into a single valued plot of particle radius R(θ) vs. polar angle θ using the CG as the reference point. One takes the particle silhouette, finds the center of gravity, selects points on the edge of the particle silhouette, runs radius vector from CG to particle edge, explodes edge segments, and finally strings the segments together in a rectalinear plot of R(θ) vs. θ. As the number of edge segments becomes more numerous the plot becomes more accurate. The R vs. θ plot can be approximated by a Fourier series.

microdata. In many analyses, only the first 10 to 20 coefficients are utilized, It is in this macroarea that most of the visual or particle recognition informaiton lies.

Besides Fourier there are other orthonormal functions that can be used to analyze the particle's silhouette. Among the more familiar are the Walsh and the Haar (Figures 6 and 7). The choice of which orthonormal functions to use depends on the particle shapes being analyzed as well as the speed of computer analysis. For a while, it was thought that the Walsh or the integral of the Walsh, which are triangular functions, would more nearly approximate that of sharp particles. Furthermore, the Walsh and similiar functions, particularly when used with Hadmard transforms, allow for the use of very fast alogrithm. For example, the Walsh functions require multiplying by plus or minus and the use of the Hadmard transforms give very sparse matrices meaning

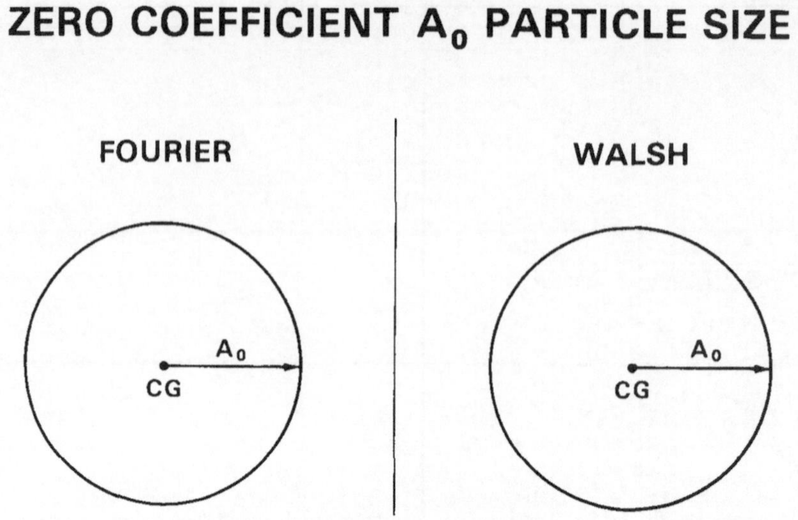

FIGURE 2. Zero coefficient when plotted in polar coordinates yields a circle for both Fourier and Walsh functions. The subtle implications is that all particles are basically spheres. Higher-order coefficients merely show deviation from sphericity. It is possible to find another idealized shape such as a cube and measure deviation from that shape, but it cannot be done with existing orthonormal functions without preparation. .

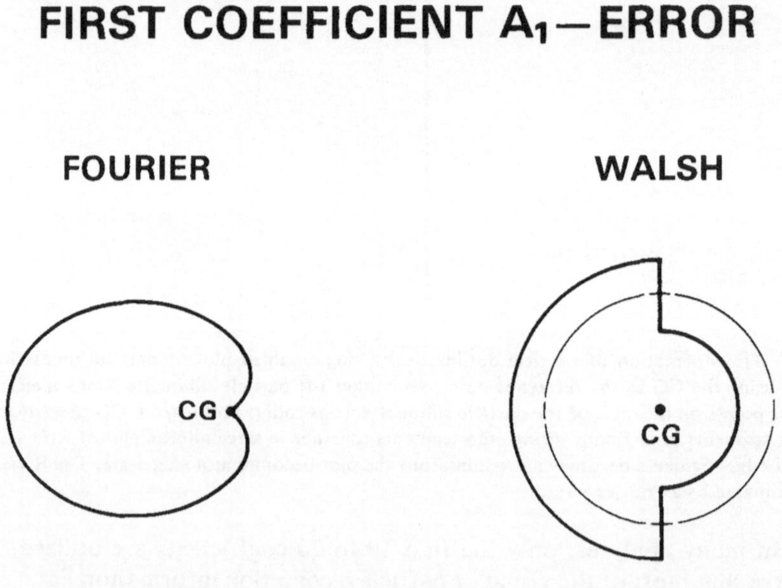

FIGURE 3. The first coefficient A_1 is not symmetrical about the CG as a plot of $A_0 + 0.2A_1$ shows. Thus the value of A_1 is the error in locating the CG correctly.

that the amount of arithmetic that needs to be done is very small. Recent developments in computors, which reduce multiplication operations to two cycles, allow for parallel processing to handle complex numbers have eliminated this objection. Furthermore, the fast Fourier transforms have made the computations extremely fast even for large matrices. As a result, all investigators are currently using Fourier techniques.

Each of the orthonormal functions yields a series of coefficients. These coefficients are linearly related to the coefficients of every other orthonormal function. From a

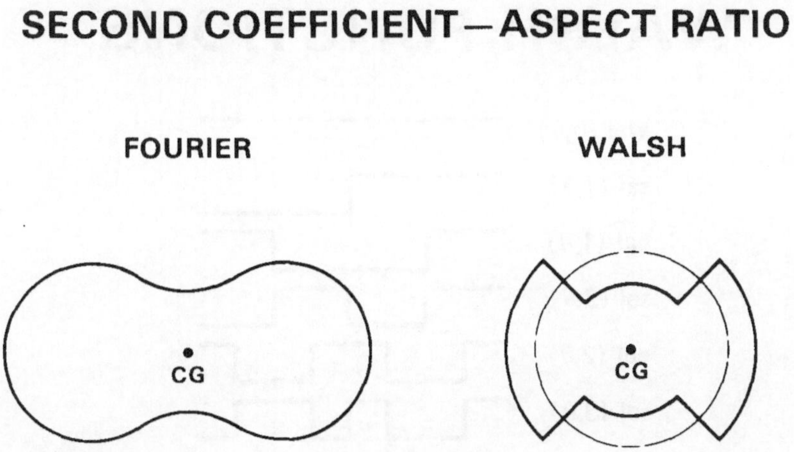

FIGURE 4. A plot of $A_0 + 0.2A_2$ showing the bilateral symmetry which all even coefficients impart to a particle. A_2 is a measure of the particle elongation or aspect ratio, and when A_2 and A_4 have similar phase angles, the particle aspect ratio may be increased.

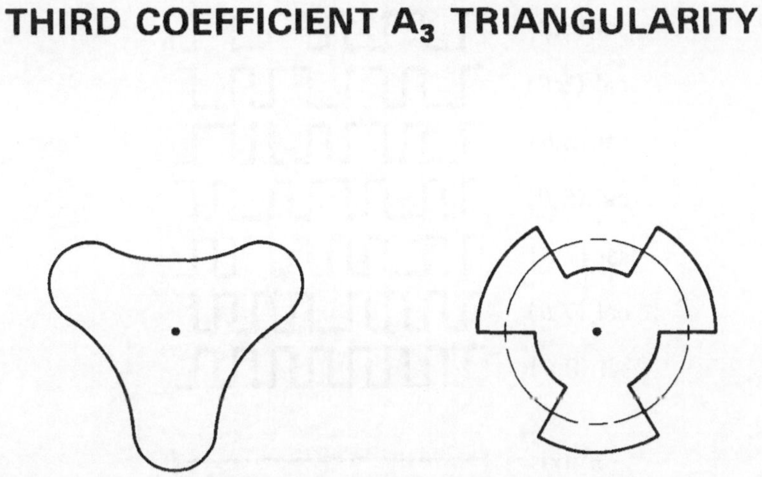

FIGURE 5. Combined plot of $A_0 + A_3$ showing the triangularity of the particle. Note the lack of bilateral symmetry imparted by all odd coefficients.

physical picture, as well as mathematical purity, the best orthonormal set to work with are the Fourier functions.

As described earlier, only certain coefficients may be needed in describing a given type of particle. The Karhunun-Loève[8] yields the optimal post-priori coefficients for any orthonormal set that need to be found. This approach is very elegant mathematically, but currently has not been employed. Its major use will probably be in specifications using a combination of coefficients to describe certain functions of the particle which are desirable and others which are undesirable.

There are several methods for obtaining the coefficients, and the coefficients thus obtained depend on the method utilized. The classical approach used by most investigators (described in Figure 1) is to find the center of the area, run a radius vector out to the perimeter of the silhoutte, and describe the magnitude of the radius vector as a function of the angle from a reference vector. The coefficients are independent, that

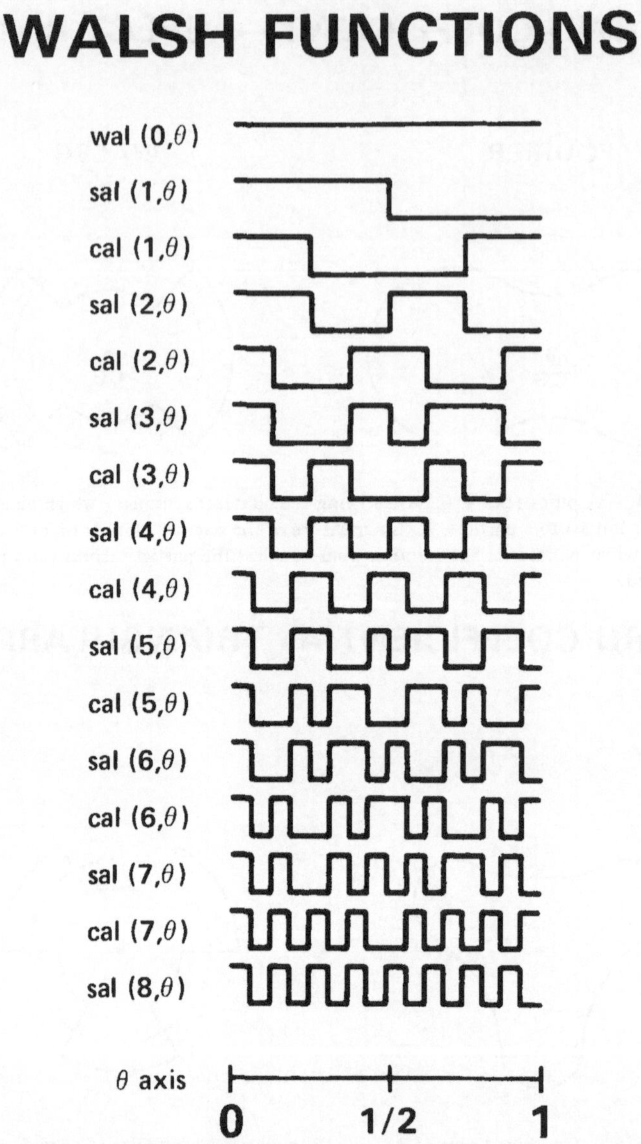

FIGURE 6. Plot of the first 16 Walsh functions which is a set of complete orthonormal functions. Note the zero crossings and both the symmetry or inverted symmetry about the midpoint. Walsh functions are about half as efficient as Fourier functions in approximating a given curve. Walsh functions have values of 1 or −1.

is to say, the magnitude of any coefficient is independent of the position or initial reference point for measurement. The Zahn[9] method used by Fong et al.[10] starts on the perimeter, goes the discrete distance along the perimeter, and measures the radius vector and change in angle from point A to point B. In other words, one starts at the perimeter at point A, goes to point B, and measures the difference of the tangent between the two systems. The advantage of this technique over the one initially described is that particles with invagination or reentries can be analyzed without difficutly. The first technique described cannot be used on more complex particle shapes because of the multivalue problem. As a result, the Zahn[9] method is used on certain

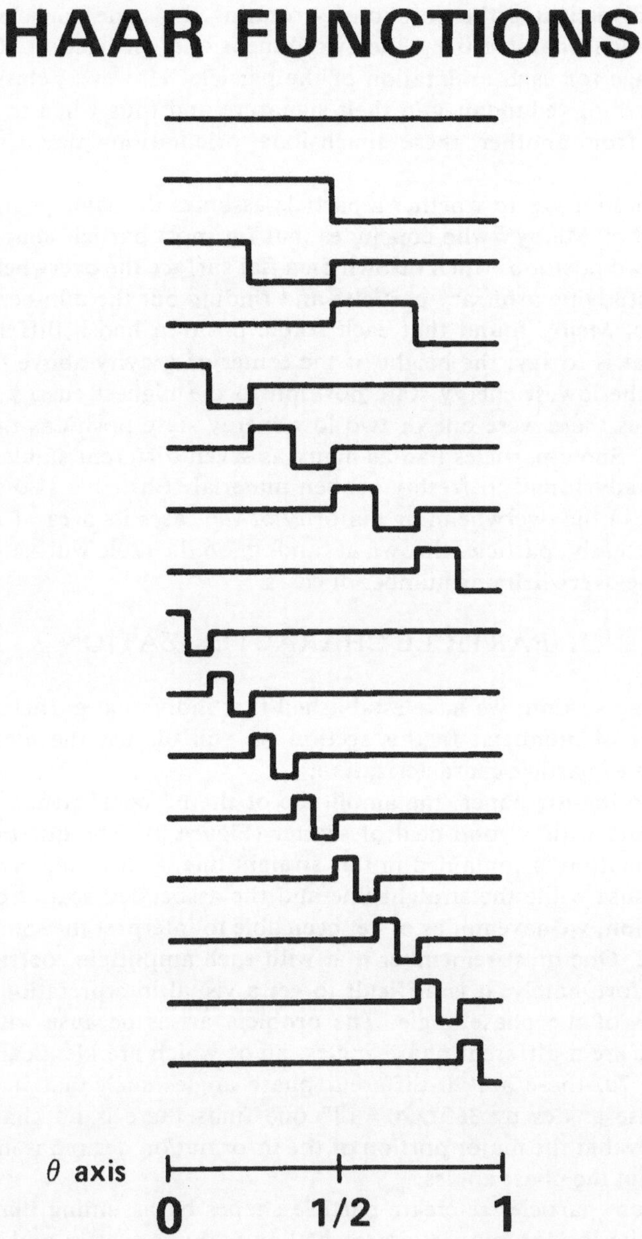

FIGURE 7. Plot of first four levels of Haar function, another orthonormal set of functions having values of 1, 0, or −1. Note that the fourth level has eight functions. The number of Haar functions increases very rapidly with the level of the function.

problems. Once again there is a linear relationship between the coefficients of the two methods.

In the method of characterization discussed so far, the particle outline or silhouette has been characterized in two dimensions. Two questions arise. First, does the particle assume a number of different positions and have a different set of characteristics for each position? Second, does the particle always assume the same position?

Ehrlich[11] has investigated the question by rotating the same particle through arbitrary axes and then characterizing the two-dimensional projection. Obviously, the coefficients change for each orientation of the particle. However, classes of particles have a high degree of redundancy in their signature and thus when trying to tell one class of particle from another, these dimensional orientations play a relatively small role.

The second question, as to whether a particle assumes the same position or not has been investigated by Meloy[12] who concludes that for most particle shapes, the particle assumes one of two positions when thrown on a flat surface the overwhelming majority of the time. By studying arbitrary particles and finding out the number of positions a particle can take, Meloy found that each stable position had a different associated energy level. That is to say, the height of the center of gravity above the flat surface increased from the lowest energy state position to the highest energy state position. For most particles there were one or two low-energy state positions that the particle usually assumed. Some particles had as many as seven different stable positions, but these were generally found in freshly broken material. Ehrlich[11] also found that the particle assumed in the overwhelming majority of the cases its area of maximum projections. Consequently, particles thrown at random on the table will asume their lowest energy state in the overwhelming number of cases.

III. PARTICLE CHARACTERIZATION

In the preceding section, we have established that individual particles can be quantified by a series of numbers. In this section we will discuss the meaning of those numbers in terms of particle characterization.

If one plots on log-log paper, the amplitude of the nth coefficient versus n, then a straight line occurs with a good deal of scatter (Figure 8). The question arises as to how much information is contained in this straight line. Unfortunately, the answer is paradoxical because while the straight line and the associated scatter contain a great deal of information, we have not as of yet been able to interpret the signatures in terms of particle shape. One must remember that with each amplitude coefficient there is a phase angle. Unfortunately, it is difficult to get a visual interpretation, that is a plot of the signatures of the phase angle. The problem arises because with an nth order coefficient, there are n different phase angles, all of which are identical. For example, with n equal to 20, there are 20 different phase angles such that if one rotates or changes the phase angles by $360°/20 = 18°$ one finds there is no change in particle shape. One hopes that the major portion of the information desired is in the amplitude coefficient and not the phase angles.

Type I[13] random particles re-create particle shapes by assuming that signatures of the particle on the log-log plot is a straight line without scatter and that the phase angle is completely random. Equations defining Type I and Type II are in Figure 9. The resultant particles in Figure 10 look, at times, like real-world particles, but they contain no surface roughness. Type II random particles assume that there is a scatter about the straight line on the log-log plot as well as random phase angles. These particles look very much like Type I random particles and they are certainly not representative of the major classes of particles such as sand grain, crushed rock, stream bed particles, etc. Type III random particles assume there is locking of the phase angles, i.e., the phase angles tend to cluster about certain points.

Odd and evenness means that the odd coefficients are consistently larger than the even coefficients for odd particles; the reverse is true for even particles. Even coefficients tend to make the particle symmetrical in that a completely even particle — all of the odd coefficients are zero — can be cut in half and rotated about the center of

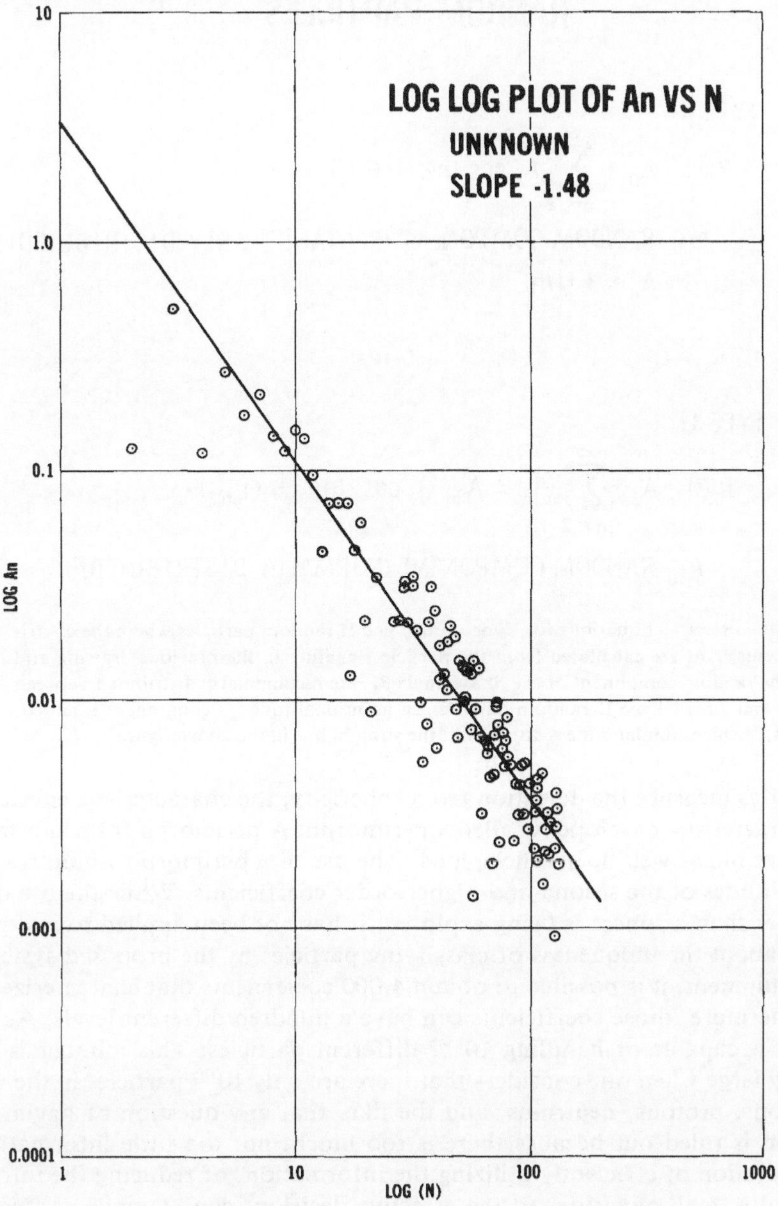

FIGURE 8. A particle signature is the log-log plot of the amplitude of the nth coefficient vs. n. Note the scatter about the line. Unplottable in a rational manner is the phase angle associated with each point. The problem arises in plotting phase angles because the tenth coefficient has ten phase angles and the nth coefficient has n coefficients.

gravity so that the two halves fall on top of each other (Figure 4). Odd particles, on the other hand, are asymmetrical (Figure 5).

Currently, investigation is going on as to the meaning of particle signatures with the eventual thought that a few simple characteristics may be able to define a class of particles quickly and efficiently.

It is interesting to note that basically the method using Fourier, Walsh, and Haar functions all assume that the particle is a sphere or that its projection, a circle, and

RANDOM PARTICLES

TYPE I

$$R(\Theta) = A_0 + \sum_{n=2}^{\infty} A_n \cos(n\Theta + \Theta_{Rn})$$

Θ_{Rn} RANDOM COMPONENT (RECTANGULARLY DISTRIBUTED)

$A_n = f(1/n)$

TYPE II

$$R(\Theta) = A_0 + \sum_{n=2}^{\infty} (\bar{A}_n + A_{Rn}) \cos(n\Theta + \Theta_{Rn})$$

A_{Rn} RANDOM COMPONENT (NORMALLY DISTRIBUTED)

FIGURE 9. Equations for Type I and Type II random particles where the coefficients, An, are calculated from the particle signature in the previous drawing and the random component of the phase angle θ_{Rn} are rectangularly distributed between θ and 2π. In Type II random particles an additional random component is added, A_{Rn}, which simulates the scatter about the straight line in the above figure.

these methods measure the deviation from sphericity, the characteristic envelope. Any simple characteristic envelope is called a perimorph. A perimorph for a substance that had cleavage might well be parallelepiped. The use of a perimorph would result in far lower amplitudes of the second and higher-order coefficients. While the use of a perimorph other than a sphere is being explored, it has not been applied to real particles.

A word about the uniqueness of classifying particles by the proposed system. With existing equipment, it is possible to obtain 4,000 coefficients that characterize the particle. Furthermore, those coefficients can have a hundred different levels. As a result, the system is capable of handling 10^{8000} different particles. This number is so overwhelmingly large when one considers that there are only 10^{100} particles in the universe, i.e., electrons, protons, neutrinos, and the like, that any question of having enough information is ruled out because there is too much, not too little information. It is merely a question of efficiently utilizing the information, of reducing the information to sufficiently small quantities so the effective decisions can be made. Actually what has happened is in the last decade we have gone from having too little information about particles to now having too much.

There are other methods of characterizing particulate matter that have been proposed. One method proposed by Kaye[14] consists of rotating the particle or its silhouette in front of a matrix of sensors and analyzing variations in the sensors' output as the particle rotates interrupting light that falls on the sensors. The use of fractals is another method proposed by Kaye[15] and yields an additional coefficient describing the particle.

IV. POWDER SIGNATURES

Up to this point we have been discussing individual particle signatures. Most industrial and scientific problems involve not individual particles, but powders. Powders

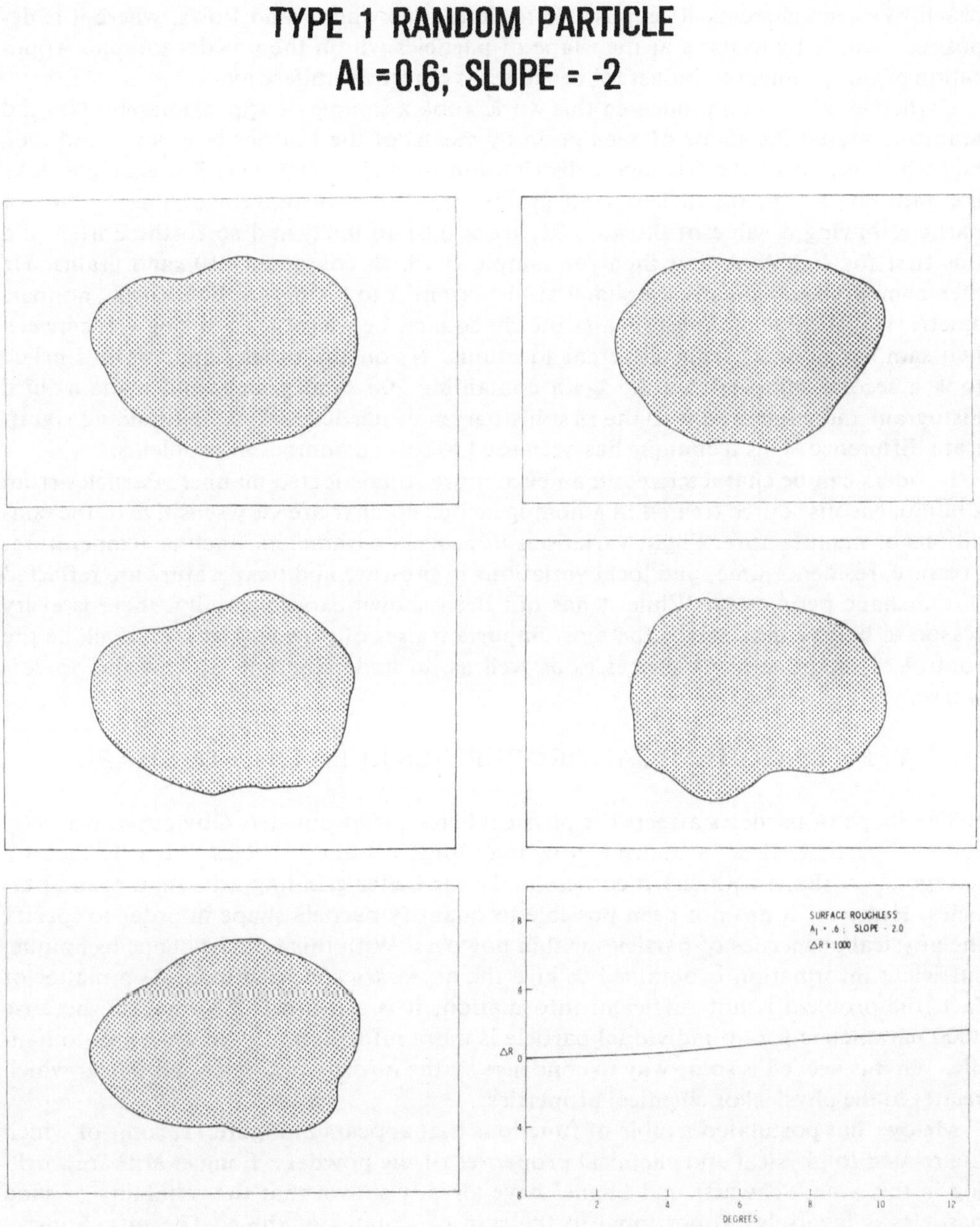

FIGURE 10. Plot of a family of random particles whose signature line had a slope of −2 and an intercept of 0.6. The particles resemble each other, have smooth surfaces — as may be seen from the expanded plot of the surface in the lower right — and look real. However, Type I random particles are expected to be very rare in the real world.

are made up of a large number of particles often from a homogeneous source. The questions arise, can one tell the difference between two different batches of powders made by similar but slightly different procedures. The work of Ehrlich et al.[16] answers this question with an overwhelming affirmative reply.

It is possible, for example, to determine whether the beach sand is coming from the dunes behind the beach or from a nearby river with the sand being washed along the beach by ocean currents. It is possible to determine where sand flows, where it is deposited, simply by looking at the shape of particles within the powder sample. Application of this problem to industrial processes is currently underway.

Ehrlich et. al.,[16] who pioneered this work, took a sample of approximately 200 sand grains, analyzed the shape of each grain by means of the Fourier technique, and then made histograms of the frequency distribution of each coefficient. For example, take the 14th coefficient of all 200 sand grains and plot a histogram of the number of particles having a value of 0.0 to 0.01, from 0.01 to 0.02, and so forth. Ehrlich did this first for sample A and then for sample B which contained 210 sand grains. He then compared statistically the shape of histogram A to histogram B. By using nonparametric statistical techniques such as the chi-square, he was able to distinguish between two samples of sand from different locations. If, on the other hand, when Ehrlich took a second sample from A, again containing 200 sand grains, and made a third histogram and compared it to the first histogram of particles of A, he found no significant difference. This technique has been used to solve a number of problems.

Powders can be characterized in an even more sophisticated manner. Particles from a homogeneous source treated in a homogeneous manner are very sensitive to the conditions of manufacture. Slight variations in process parameters, such as temperature, pressure, residence time, and local variations in pressure and temperature are reflected in the shape perimeters. While it has not been shown experimentally, there is every reason to believe that one of the most important uses of powder signatures will be the control of manufacturing processes as well as an understanding of how the process works.

V. PHYSIOCHEMICAL PROPERTIES RELATING TO SHAPE

The shape of particles affects the physical behavior of powder. Obviously, perfectly rounded particles flow far more readily than long, eel-like particles. Particles that are tetragonal in shape with sharp corners make far better grinding grits than round particles. Hitherto, it has not been possible to quantify particle shape in order to specify the physical properties of particles within powders. With this Fourier-shape technique, sufficient information is obtained to give the necessary specifications. As a matter of fact, the problem is not sufficient information, it is too much information, because 4000 parameters for an individual particle is more information than one cares to handle. What is needed is some way to condense all the information into a perimeter which relates to the physical or chemical properties.

Meloy[17] has postulated a table of functions that appears in Figure 11, some of which are related to physical and chemical properties of the powders. Fong et al.[10] are working in this area. Schwartz and Shane[1] have already shown that the rollability of sand particles is inversely proportional to the sum of squares of the coefficients. Fong et al.[10] have shown that as powders are dissolved in acid, the smoothing of the particles is a function of the sum of the coefficients. It may be that such things as both initial and final packing density, angle of repose, sinterability, chemical reactivity, and a variety of other physical properties may be related to the functions proposed by Meloy.

Because chemical bonds are broken when surface edges and corners are made, these areas are more reactive both physically and chemically. Therefore, a measure of the angle and length of edges, the number and solid angle of corners would predict the reactivity of powders.

It is not futuristic thinking to expect in the near future that abrasives, pottery and ceramic clay, metal powders for sintering, and even food such as grains will be purchased on the basis of chemical and shape specifications. These specifications may well be based on the functions shown in Figure 11.

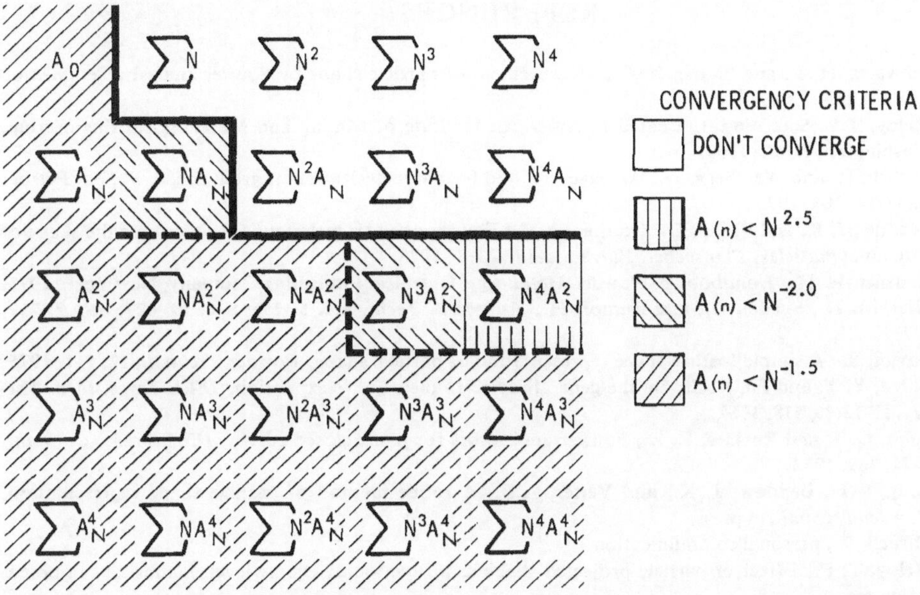

FIGURE 11. Table 2 is a compilation of 25 functions some of which in the hatched area are convergent and are believed to have physical significance. For example, Schwarcz and Shane[1] have shown the ΣAn^2 is a measure of the rollability of sand.

VI. INSTRUMENTATION AND EXPERIMENTAL TECHNIQUE

Initially, to obtain the information for the Fourier analysis, each sand grain was measured by hand. Since the average sample contained between 100 and 200 samples, the initial research was tedious and time-consuming. The latter half of the 20th century is called the age of instrumentation and it is no different with particle-shape analysis. Early production runs on the sand took 4 min per particle, but, for example, the modern Pagan Mark I[18] instrument is capable of handling 30 grains per second.

The computations which go along with the interface data are equally fast and can keep up with the instrument taking data. Thus it appears at the present moment that a sample of 200 particles can be analyzed in a little under 7 sec. Thus it is not data taking but sample preparations that is currently the limiting factor. In the future, dedicated instruments can be expected to increase this speed.

VII. SUMMARY

A sophisticated, high-speed technique for quantifying particle shapes has been developed and used. The method measures numbers such that the particle shape may be reproduced beyond any errors specified. Furthermore, different samples of powders can be compared and shown whether they are a member of a given specimen or not. In addition, physical and physiochemical properties of the powders can be obtained or predicted from the shape analysis. While the technique has raised many deep theoretical questions, it has also been very useful in solving practical problems.

REFERENCES

1. **Schwartz, H. P. and Shane, K. C.**, Measurement of particle shape by Fourier analysis, *Sedimentology*, 13, 213, 1969.
2. **Meloy, T.P.**,Screening, presented to American Institute of Mining and Metallurgical Engineering, Washington, D. C., 1969.
3. **Ehrlich, R. and Weinberg, G.**, An exact method for characterization of grain shape, *J. Sed. Petrol.*, 40(1), 11, 205, 1970.
4. **Beddow, J. K. and Philip, G.**, On the use of a Fourier analysis technique for describing the shape of individual particles, *Planseeber. Pulvermetall.*, 23, 1, 1975.
5. **Hausner, H. H.**, *Handbook of Powder Metallurgy*, Chemical Publishing Company, New York, 1975.
6. **Phal, M. H., Shadel, G., and Rumpf, H.**, *Aufbereit, Tech.*, Sec. 5, 257; Sec. 10, 627; Sec. 2, 759, 1973.
7. **Davies, R.**, A simple feature-space representation of particle shape, *Powder Technol.*, 12, 111, 1975.
8. **Chien, Y. T. and Fu, K. S.**, On the generalized Karhunen-Loève expansion, *IEEE Trans. Info. Theory*, IT-13(3), 518, 1967.
9. **Zahn, C. J. and Roskies, R. Z.**, Fourier descriptors for plane closed curves, *IEEE Trans. Comput.*, C-21, 269, 1972.
10. **Fong, S-T., Beddow, J. K., and Vetter, A.F.**, A refine method of particle shape representation, *Powder Technol.*, in press.
11. **Ehrlich, R.**, personal communication.
12. **Meloy, T. P.**, Effect of particle projected shape as a function of energy level, manuscript in preparation.
13. **Meloy, T.P.**, Type I random particles, Proc. Fine Particle Soc., Int. Powder and Bulk Solids Conf., Chicago, in press.
14. **Naylor, A.G., Robin, G.A., and Kaye, B.H.**, Shape analysis of ceramic particle profile from electron and photo micrographs, *Fine Particles, Second International Conference*, Kuhn, W.E., Ed., Electrochemical Society, 1974, 28.
15. **Kaye, B. H.**, Specification of the ruggedness and/or feature of a fine particle profile by its fractal dimension, *Powder Technol.*, in press.
16. **Ehrlich, R., Orzeck, J. J., Weinberg, B.**, Detrital quartz as a natural tracer — Fourier grain shape analysis, *J. Sed. Petrol*, 44(1), 145, 1974.
17. **Meloy, T. P.**, Particle shape characterization — recent developments, in *Testing and Characterization of Powders and Fine Particles*, Beddow, J. K., and Meloy, T. P., Eds., Heyden and Son, Ltd., London, in press.
18. **Ehrlich, R.**, personal communication, 1978.

Chapter 3

ANALYSIS OF PARTICLE MORPHOLOGY DATA

Robert Ehrlich, P. Jeffrey Brown, Jeffrey M. Yarus, and Duane T. Eppler*

TABLE OF CONTENTS

* Present affiliations: P. Jeffrey Brown, Exploration Geologist, Mobil Oil Exploration and Producing, New Orleans; Jeffrey M. Yarus, Exploration Geologist AMOCO Production Company, New Orleans; and Duane T. Eppler, Post-doctoral Fellow, Physics Department, West Virginia University, Morgantown.

INTRODUCTION

The particle shape analysis system which we have developed over the past decade has always been directed towards problem solving, generally in the field of sedimentary geology. The system which has evolved and the generalities discussed in this section are not the result of comprehensive theoretical analysis concerning the fundamental nature of particle shape, nor the optimal mathematical expression of shape. The particles with which we have worked most are quartz grains in the sand and silt size ranges (.0039 to 2mm). Quartz grains are generally convex in shape and so represent a class of relatively simple albeit bumpy shapes. Although the specifics of the foregoing discussion will be linked to quartz particle shape data, the general approach should be applicable to most particle shapes.

A question that generally arises early in shape analysis concerns the number of measurements necessary to characterize the shape of a given particle. The first step in answering this is to determine the efficacy of the use of two-dimensional projections or silhouettes of the particle for shape analysis. Silhouettes prove adequate in two situations: (1) when all potential silhouettes of a particle can be predicted from any one silhouette of that particle (lack of independent projections); or (2) when a unique silhouette can be selected which carries sufficient information. In the case of quartz, we use a unique projection, the maximum projection profile of the particle. We and others have shown in past studies[1,2] that projections of quartz grains mounted on glass slides are almost always within 3 to 5% of their maximum projection area. Thus it is relatively simple to obtain maximum projection profiles.

Use of any projection of a three-dimensional particle, including the maximum projection profile, always carries the risk that insufficient shape information is carried in two dimensions for the problem at hand. In the case of quartz, a large number of carefully designed experiments[2-6] have shown that the maximum projection profile is relatively dense with information and that other profiles are redundant with respect to this profile. This is not to say the additional information does not lie in the integrated three-dimensional shape of a particle. However, this additional information, if any, has not been needed in the investigations we have pursued so far.

The number of measurements necessary to describe the shape of a two-dimensional profile is a function of the degree of inherent symmetry of the shape as well as the degree of precision needed. Obviously, shapes with high inherent symmetry (circles, squares, etc.) can be described by one or two variables. As shapes increasingly depart from symmetry, the number of variables needed to completely describe the shape increases. This is not to say that shapes need always be described completely. Something less, for instance the aspect ratio, may carry sufficient information with respect to a specific problem.

In sedimentology, particle shape has commonly been measured by fewer variables than needed to completely describe the shape. However, these shape measures (sphericity and roundness[7]), which have been assiduously applied to problems for more than 100 years, have conspicuously failed to shed any light upon any significant geological problem. These ineffective shape measures are precisely those that work best with regular, symmetrical shapes and lose their value as shapes depart from regularity. This is one of many cases where definition of a shape measure carries with it an implicit, unproved, a priori assumption of high regularity. At the onset of any particle-shape investigation, however, it is safest to assume the converse: that the shapes are irregular and asymmetrical, because any measure sufficient to evaluate such shapes will also be effective with more symmetrical shapes.

Thus, in the case of our quartz particle shape analysis system, we had to begin with a complete measure of maximum projection profile — completeness meaning that the empiric shape can be approximated as precisely as needed. This dictates use of a series which converges to the empiric shape. In our case, we use a discrete Fourier series in closed form,[8,9] however, any series that converges can be used. Through the avenue of a Karhunen-Loève expansion (principal component analysis), information contained in a convergent series can be transformed into any other series. Thus, selection of the converging series used (Fourier, Walsh, etc.) is based primarily on grounds of convenience.

In our quartz studies, the amplitudes of the first 20 terms (harmonics) of the Fourier Series are used to describe the maximum projection profile (Figure 1). However, in our work, as in most, our shape studies do not involve analysis of one single particle. Instead, an investigation involves thousands of particles. In the case of, say, a sample containing 200 particles, the data space would contain 4,000 values (20 amplitudes × 200 grains). The evaluation and exploration of this sort of data space is the principal subject of this paper.

The tactics of analysis of multiparticle shape data will be governed by two factors: (1) the investigational objectives; and (2) the nature of the shape frequency distribution. The general objectives for a shape measurement investigation defines the experimental design. The design specifies the sampling plan and the general families of numerical-statistical techniques which should be used in evaluating and interpreting the data. A complete analytical strategy, however, cannot be defined by the experimental design alone, but must take into account the nature of the underlying shape frequency distribution. A shape frequency distribution, like a size frequency distribution, is the proportion of shapes occupying various intervals of the shape spectrum. It is precisely the same as any statistically defined frequency distribution.[10-12]

The shape frequency distribution of two-dimensional projections differs from size frequency distributions in that size is measured univariately while shape is commonly a multivariate property. In the case of quartz, we measure the amplitudes of 20 harmonics for each shape, and thus quartz particle shape frequency distributions can in this case have as many as 20 dimensions. However, the amplitude values are not statistically independent (e.g., significant correlations exist between harmonics) so that the true dimensionality of quartz shapes is less than 20. Principal component analysis[13,14] of the collection of quartz outlines from single samples indicates that these outlines can be described by no less than six independent variables, linear combinations of which produce the 20 harmonic amplitudes. This means that at least six variables are necessary to completely describe quartz shapes and so that the shape frequency distributions for quartz can probably never be less than six dimensional. If this multivariate shape frequency distribution were multivariate normal or multivariate log-normal, the analysis of the data could follow the conventional lines outlined in any basic text on multivariate statistical analysis. However, in the case of quartz such simple distributions do not occur, and analysis must proceed along less conventional lines.

II. TACTICS OF PARTICLE SHAPE ANALYSIS — PRINCIPLES AND EXAMPLES

Particle-shape analysis cannot be an end in itself. So what are some of the applications of this technique? The diversity of shapes within a sample reflects the diversity of the process which created the shapes. Inasmuch as particle formation is generally complex and difficult to observe directly (e.g., erosion, ball milling, aerosol formation, etc.), analysis of the diversity of the final end product can yield information about the variability of process and the number of competing processes involved in particle for-

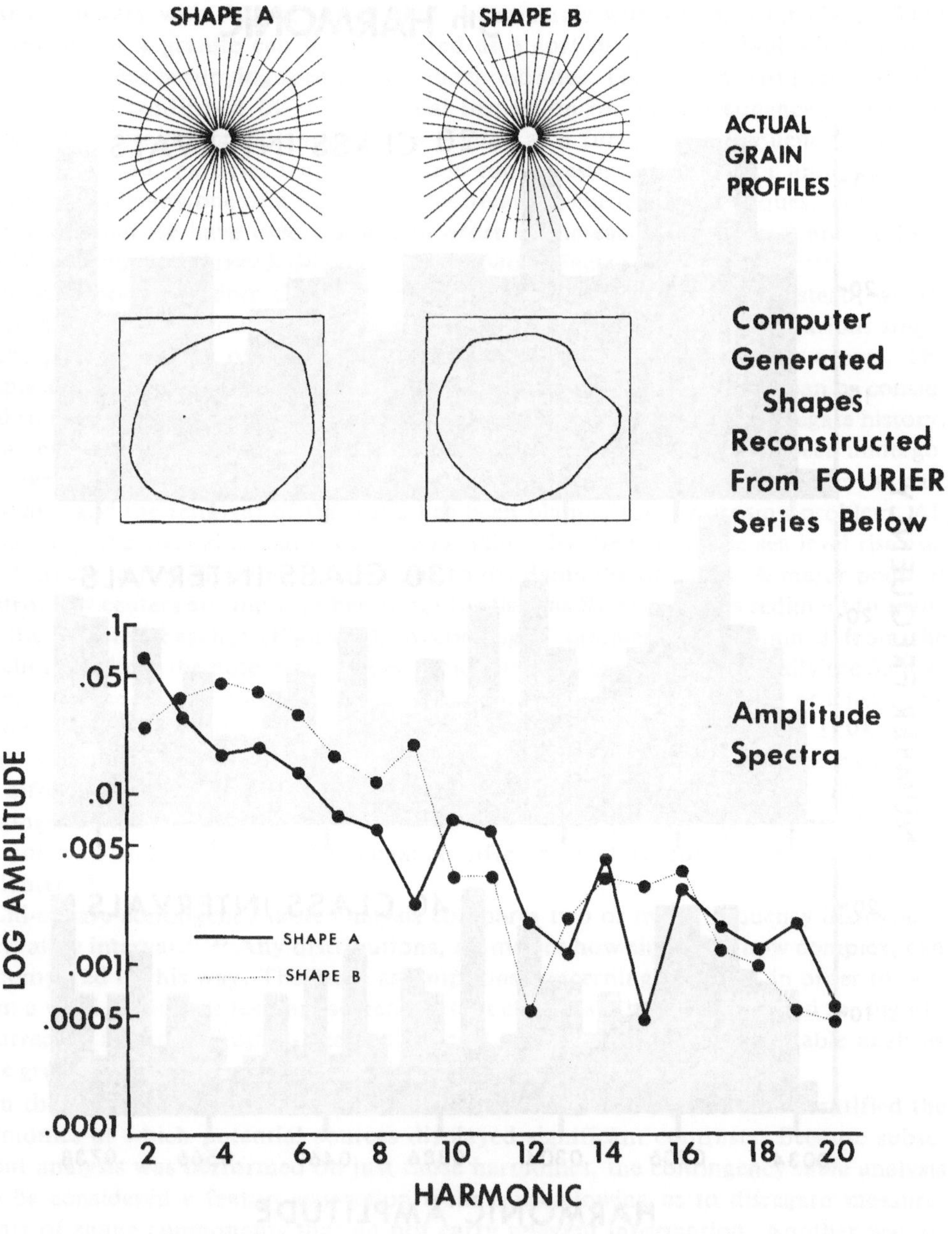

FIGURE 1. Harmonic amplitude spectra for two particles.

mation. Likewise, processes acting on particles after their formation (abrasion, solution, etc.) can be monitored and evaluated. Particle-shape analysis also bears on he aggregate behavior of solids such as trafficability, porosity and permeability, and packing. This list of applications, while far from exhaustive, reflects the main thrust of our research over the past 10 years.

In light of the previous discussion, the first task is to evaluate the nature of the shape frequency distribution. In the case of quartz, the distribution is generally complexly polymodal (Figure 2). As it is eroded from its crystalline source rock, quartz sand and silt shape frequency distributions have three or more modes. As natural

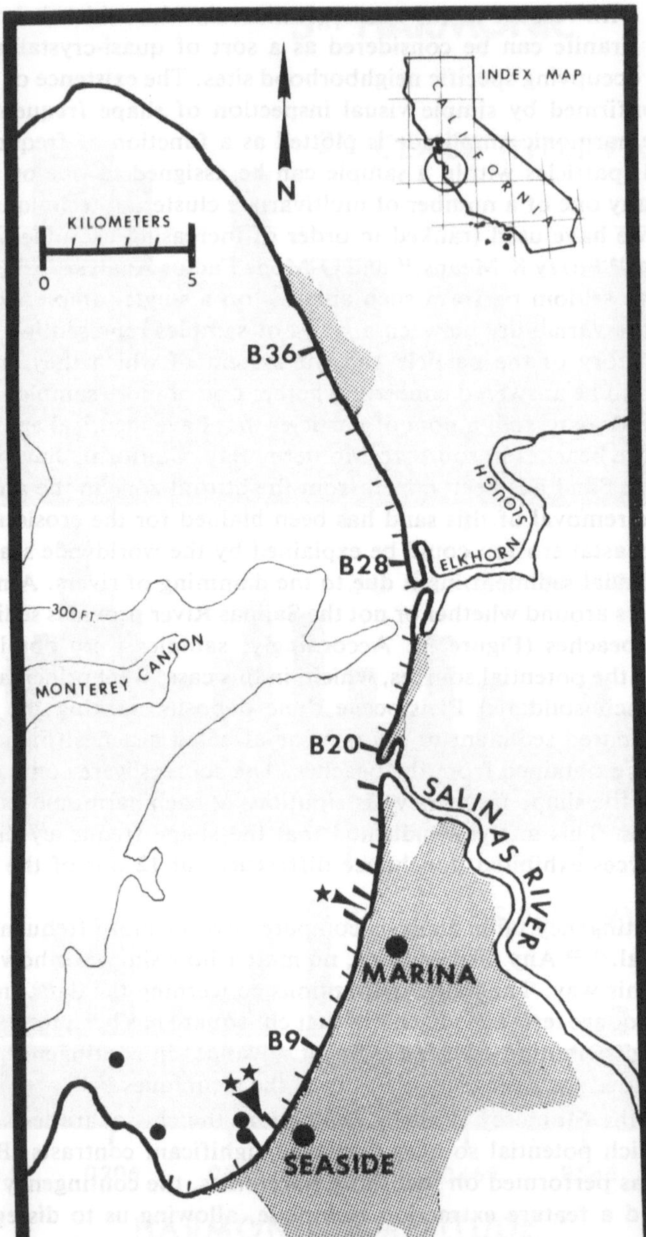

FIGURE 3. **Location of samples from Monterey Bay, California (tick marks). Mining sites (stars) located at landward tip of triangular symbol. Shaded area represents location of Pleistocene dune deposits which are a potential source. (From Porter, G. A., Ehrlich, R., Osborne, R. H., and Combellick, R. A., *J. Sed. Petrol.* in press. With permission.)**

nels. Here, as in Monterey Bay, analysis commenced with contingency table comparison of end-member samples, and the 14th harmonic was identified as one which displayed a significant contrast. In addition, at the 14th harmonic, this contrast between

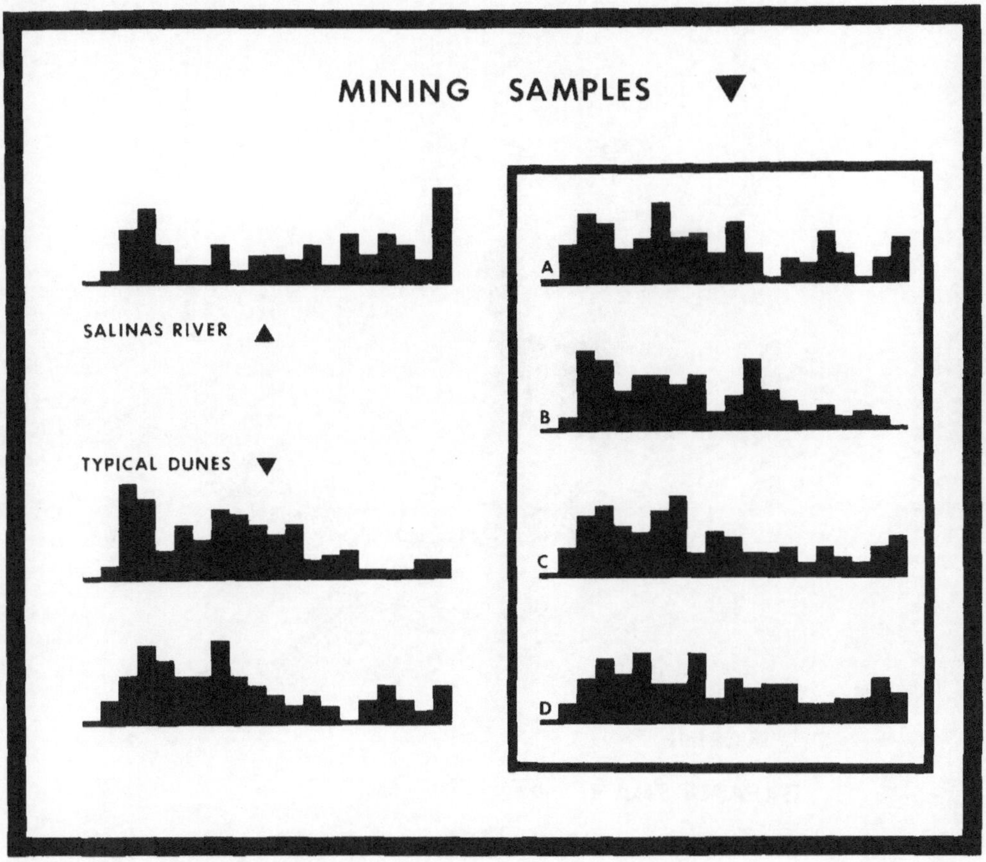

FIGURE 4. Histograms of amplitudes of Harmonic 17 comparing beach-littoral samples from mining sites with potential sources. Similarity of mining site samples is close; contrast with Salinas River sands is marked, indicating that little if any Salinas sand is being added to the southern beaches. (From Porter, G. A., Ehrlich, R., Osborne, P.H., and Combellick, R. A., *J. Sed. Petrol.*, in press. With permission.)

the littoral and Cooper River samples occurred on the right side of the distribution (Figure 6), thus allowing relative proportions to be determined by calculating the proportion of grains in that interval (Figure 7). When each sample was evaluated in this way and the proportion of the marine component plotted on the map (Figure 5), it showed that the marine sands are dominant in the dredged channel up to 15 km inland.[25] The Charleston Harbor investigation was the first in which we successfully unmixed samples into their constituent components. However, the technique used was considered to be generally unsatisfactory because it required the existence of end-member samples as well as great contrasts on the tails of one of the shape frequency histograms. Subsequent studies use a more general technique, Q-mode factor analysis.[13,17-19] This technique can determine the relative proportion of components and characterize end members even if they are not present in the sample array. Unfortunately, Q-mode factor analysis carries with it a heavy burden of jargon and a long history of misapplications. The reader is cautioned, therefore, to apply it with care.

A recent investigation of sand transport patterns in the Gulf of Alaska illustrates the usefulness of the technique. Major quantities of terrestrially derived sand are introduced to the shelf through only three river and glacier systems in the area (Figure 8). However, these sources pour enormous volumes of sediment into the Gulf — roughly

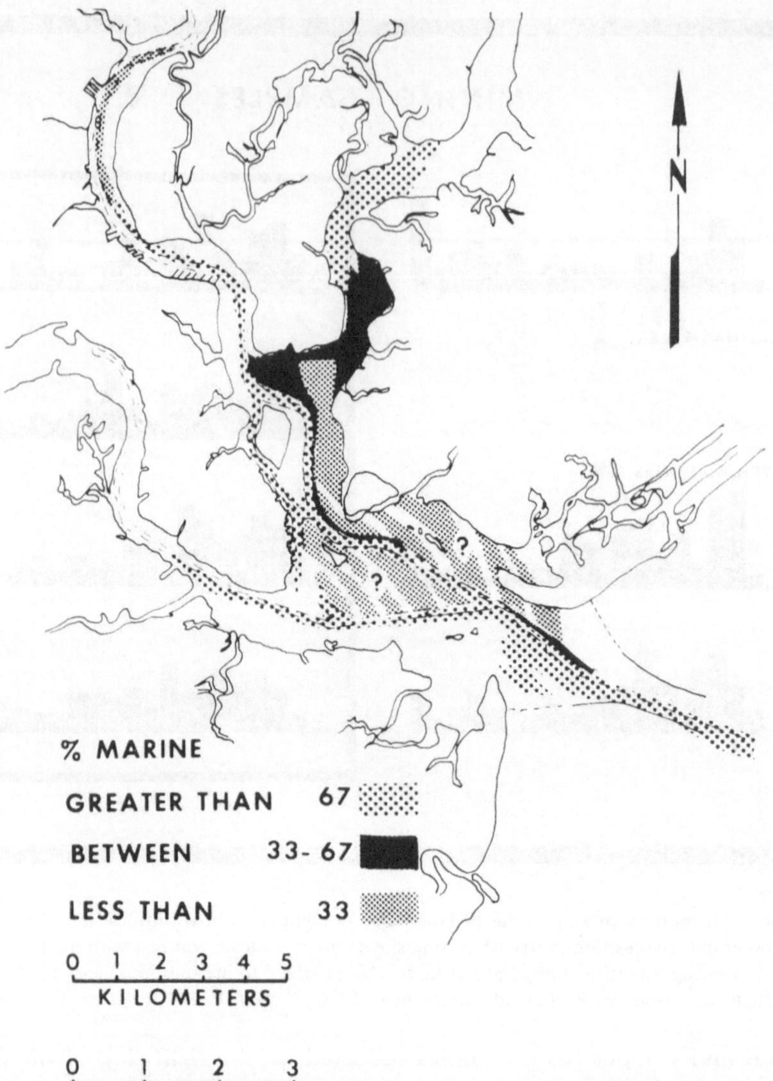

FIGURE 5. Location map of Charleston Harbor, South Carolina. Proportion of marine sand in the sand fraction shows dominance of marine sands in the harbor channels. (From Van Niewenhuise, D. S., Yarus, J. M., Przygocki, R. S., and Ehrlich, R., *J. Sed. Petrol.*, 48, 373, 1978. With permission.)

2 million metric tons a year. The area has the highest winter cyclone frequency in the Northern Hemisphere as well as an extremely large tidal range. The combination of large sediment input from few sources and rigorous wave and tidal regime provide optimal conditions for sediment transport, making the Gulf of Alaska an ideal place to test shelf transport models.[26]

Part, if not all, of the sedimentary system of the Eastern Gulf of Alaska is very active. Large amounts of sediment issue from a limited number of sources into the Gulf, and move and mix in at least the shallower portions. Some, possibly all, the differences between samples reflected by significant chi-square results may be due to differences between samples consisting of various proportions of mixtures from these

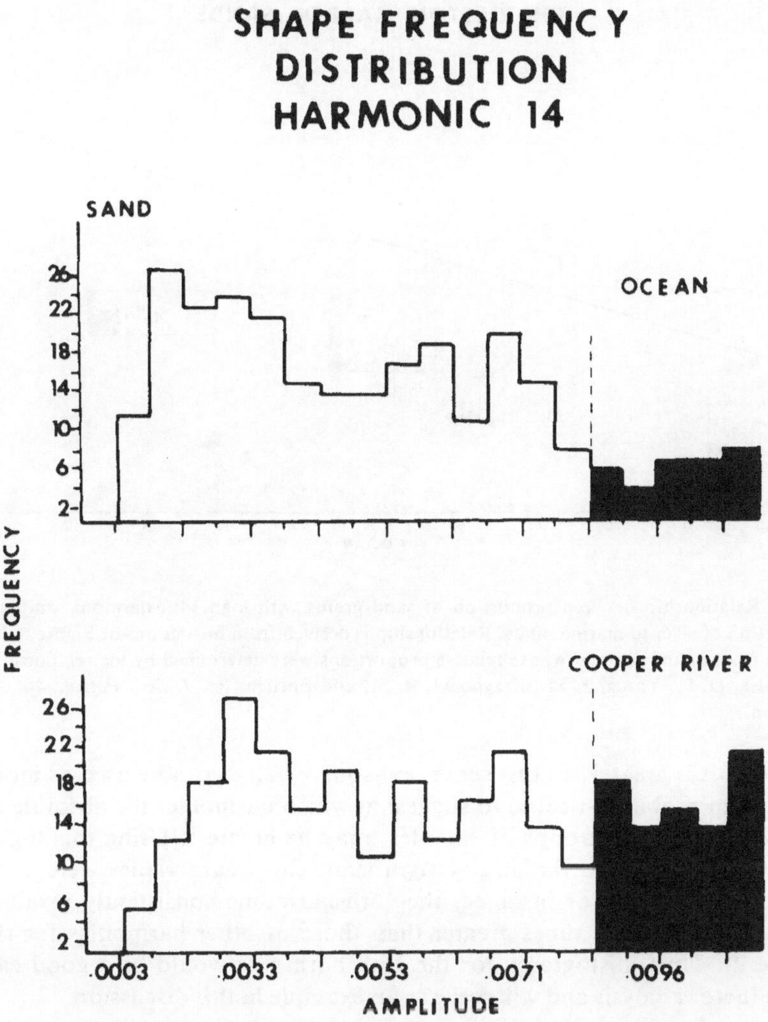

FIGURE 6. Histograms of amplitudes of the 14th harmonic illustrating the difference between fluvial and marine sands. (From Van Niewenhuise, D. S., Yarus, J. M., Przygocki, R. S., and Ehrlich, R., *J. Sed. Petrol.*, 48, 373, 1978. With permission.)

sources. If so, then gradients with respect to quartz types may exist in the region. For instance, the Bering River may generate distinctively shaped quartz grains which mix and are progressively diluted in the littoral zone. Chi-square, however, is not well suited to detect such a gradient. Documentation of significant differences between potential and members with chi-square justifies use of another technique, Q-mode factor analysis,[17] to detect relative proportions of end members in such natural mixtures. Chi-square results are used as a guide for application of factor analysis.

Absolute chi-square values will be used in that light as an indicator of the degree of contrast between samples, in addition to using chi-square in a probabilistic sense. Although any two given chi-square values may exceed a critical tabular value, it is not correct to say that the greater of the two is more significant than the other. Both results are equally unlikely to have arisen by chance. However, if one chi-square value is markedly larger than the other, this can indicate that the nonchance sample difference

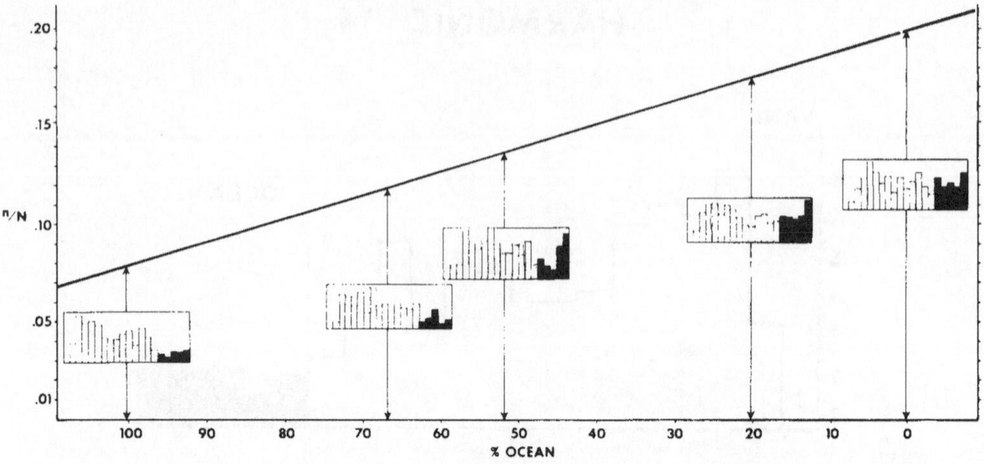

FIGURE 7. Relationship between proportion of sand grains with high 14th harmonic amplitudes and relative proportion of river to marine sands. Relationship is derived from histograms of Figure 6. Illustrated histograms are from actual samples whose relative proportions were determined by the relationship. (From Van Niewenhuise, D. S., Yarus, J. M., Przygocki, R. S., and Ehrlich, R., *J. Sed. Petrol.*, 48, 373, 1978. With permission.)

associated with it is greater. In this sense, chi-square values can be used as index numbers (rather than probabilistically) to suggest at which harmonics the absolute contrast between two samples, or groups of samples, may be greatest. Using this logic, those harmonics which generated the largest significant chi-square values were chosen for Q-mode factor analysis. For instance, the 16th harmonic consistently produced chi-square values two to three times greater than those of other harmonics for the same tests (Figure 9). Thus, histograms for the 16th harmonic would be a good candidate for Q-mode factor analysis and will serve as an example in this discussion.

On the basis of previous work[27] it is expected that each source will tend to dominate specific regions of the shape frequency distributions, so that increasing proportions of any source in any sample will tend to inflate specific portions of the sample frequency distribution.

The variables used in Q-mode factor analysis are the relative proportions of grains in each category of the shape frequency histograms at a given harmonic. For instance, given an array of samples, each amplitude frequency histogram for the 16th harmonic was divided into 20 intervals. The relative proportion of grains in each interval was determined (Figure 10). From the data matrix, similarity coefficients between all pairs of samples were determined, thus yielding a similarity matrix.[13] The similarity coefficient used, the cos theta coefficient is analogous to the correlation coefficient. It is this matrix that is analyzed in the factor analysis. The algorithm Extended CABFAC[18] was expressly designed for this kind of analysis, and was therefore used in this study.

Q-mode factor analysis indicated that most of the samples could be explained in terms of a two-source, or end-member model. That is, communalities for two factors exceed 0.90 for most samples, indicating that over 90% of the variation can be accounted for using two end members. As an illustration of this, Figure 11 displays a graph of communalities against the number of factors (two through five) for the 16th harmonic. The shaded region represents the zone of high communalities, where most

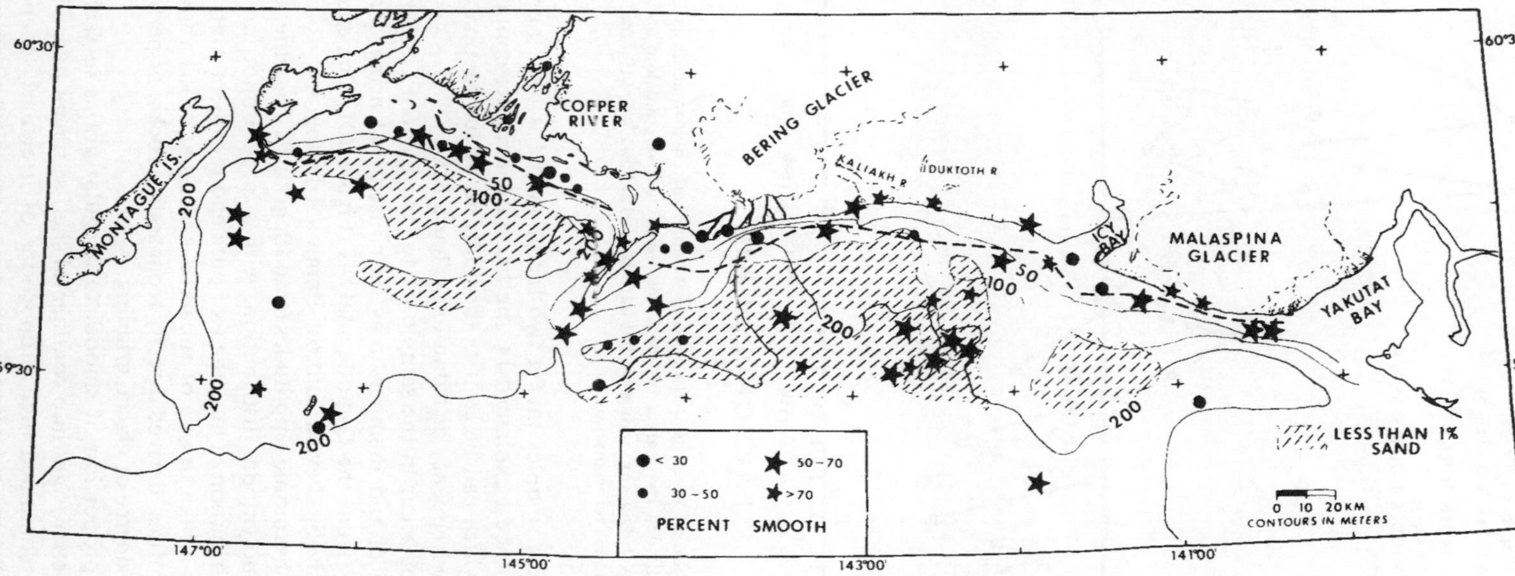

FIGURE 8. Sample locations, eastern Gulf of Alaska. Areas shown in pattern represent zones containing less than 1% sand. Dashed index line delineates two distinct zones on the inner shelf: an inner zone dominated by less than 50% smooth grains, and an outer zone dominated by samples whose proportions are greater than 50%.

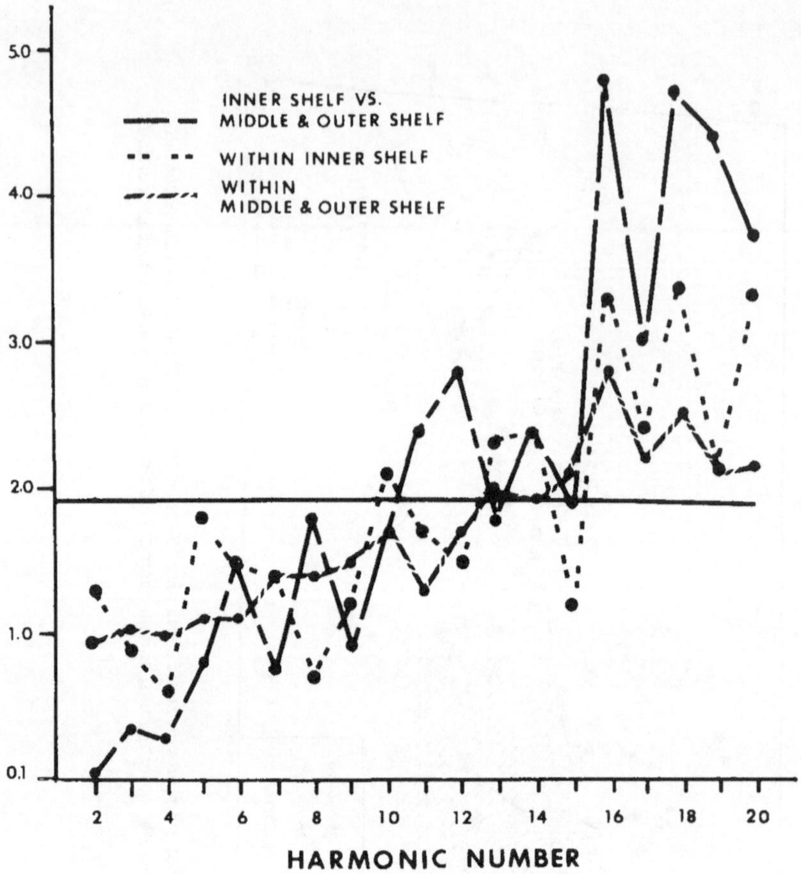

FIGURE 9. Evaluation of variation between and within shelf environments expressed a chi-square divided by degrees of freedom. Values above solid line are significant at the .05 level. Diagram indicates that harmonics 12, 16, 17, 18, 19, and 20 are optimal for Q-mode factor analysis.

samples occur. It is for these samples that a two-factor model is most appropriate. Samples which could not be best explained by the two-factor model contain one or more unique components which can be seen as the addition or subtraction of modes in their corresponding shape frequency distributions (Figure 12).

Pure end members do not occur in the sample set for the Eastern Gulf of Alaska. Histograms of the theoretical end members may be constructed using composition scores of the reference axes from the Q-mode analysis (Figure 13). One end member is characterized by a positively skewed, almost unimodal histogram, consisting of grains with dominantly low-harmonic amplitudes (smooth grains). The other end member is a more commplexly polymodal distribution, with a large proportion of grains on the right side for the distribution. Thus, the two pure end members contrast grains that have been subjected to a high degree of subaerial and subaqueous abrasion with grains of highly irregular profiles, derived directly from source rocks, or perhaps rocks which have undergne a high degree of glacial grinding.

By studying the relative proportion of smooth to irregular grains in each sample, Yarus[26] was able to gain a better isight into continental shelf sediment hydrodynamics. For instance, samples that contained high proportions of irregularly shaped grains tended to be located where prominent sources are near volcanoclastic bedrock or gla-

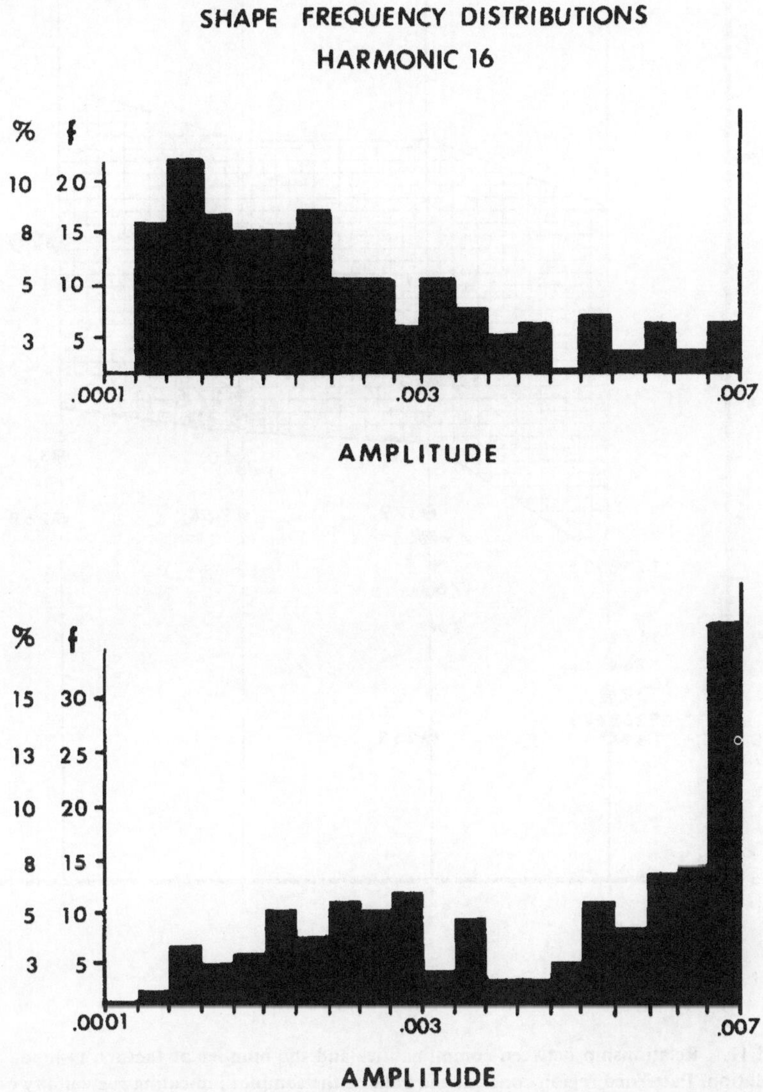

FIGURE 10. Shape frequency distributions for two samples showing percentage of grains in each interval. The percentage of grains in each of the 20 intervals are utilized as input for Q-mode factor analysis.

cially sculpted deposits. In contrast, samples with high proportions of smooth grains tend to occur in sand-poor regions of the continental shelf, such as major sea valleys and troughs.

Most importantly, Q-mode factor analysis permitted Yarus to separate the sand on the inner continental shelf into two distinct, apparently depth-controlled zones (recently derived vs. relict sand), with a boundary which runs almost parallel to the trend of the coast (Figure 8). This data provided the first actual quantitative evidence in support of the theories of Swift[28] and others which state that sand recently derived from the continental landmass cannot be transported seaward of the near-shore zone without a major fall in sea level.

Thus, for particle-morphology analysis involving mixtures, Q-mode factor analysis will probably be a standard analytical technique. Factor analysis, both R-mode and

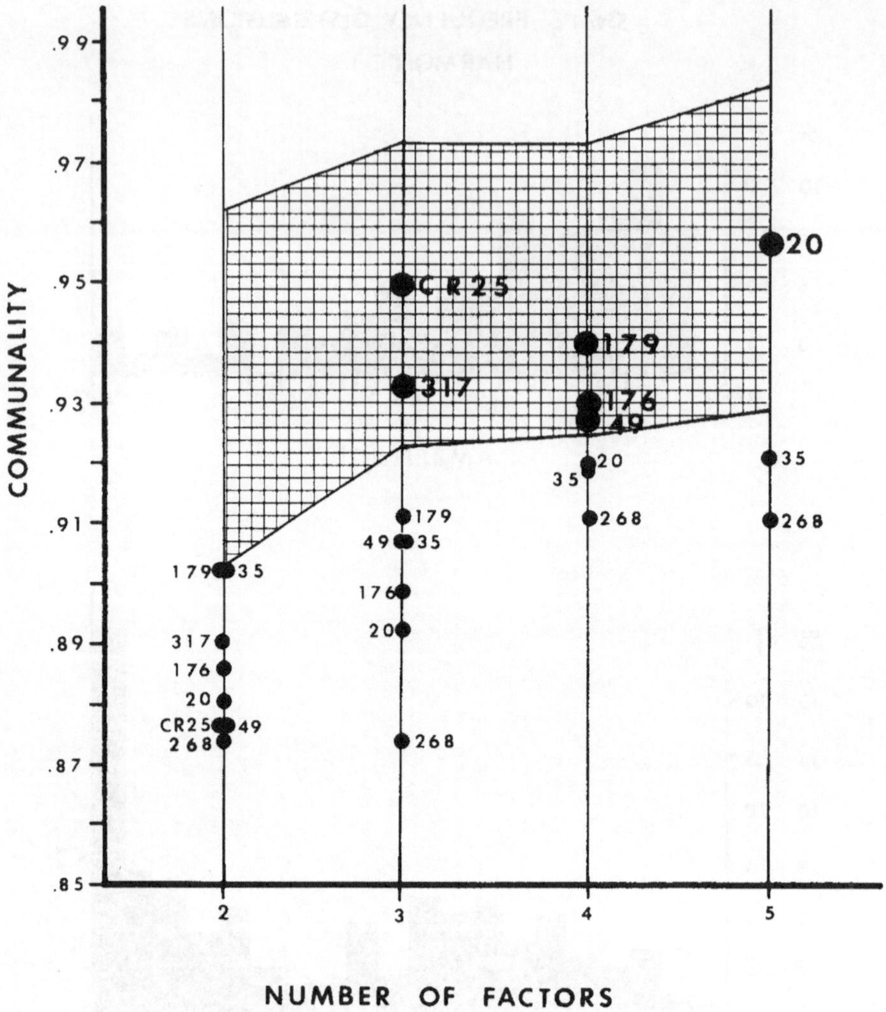

FIGURE 11. Relationship between communalities and the number of factors, Q-mode VARIMAX solution. Patterned region contains the bulk of the samples indicating the validity of a two end-member solution for most samples. Some of the outliers, shown at two factors, move into the shaded region with the addition of more factors.

Q-mode, has much broader applications than the uses mentioned here, but descriptions of such analyses are too lengthy to be included here. Because of the complexity and prolixity of much of the factor analysis literature, the authors highly recommend the lucid presentation of Davis[14] as an introduction to the technique and the papers by Klovan and Miesch[18,19] and Klovan and Imbrie[17] for further reading. Only after careful perusal of these sources should the work of Jöreskug et al.[13] be consulted.

The evolution of a shape frequency distribution under the influence of shape modification processes is another subject of great interest to us. One aspect is to change from the complex polymodal shape frequency distribution characteristic of the early stages of quartz morphology evolution to the smooth, almost unimodal, skewed distribution that characterizes highly abraded quartz (Figure 14). Does the process of transformation involve, through abrasion, destruction of modes at high amplitudes, or do the modes migrate continuously towards lower amplitudes? A similar question can be

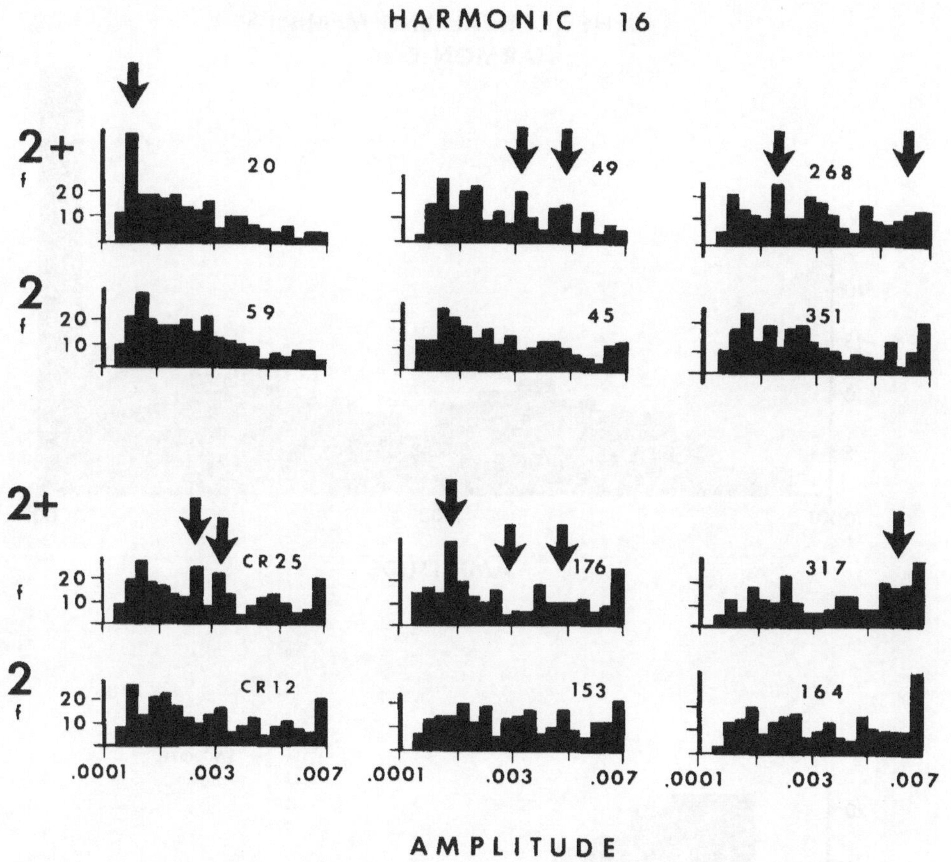

FIGURE 12. Comparison of samples containing more than two end-members (2+)with a two end-member sample (2). Both samples of each pair contain equal percentages of the smooth end member. Arrows indicate probable location of additional component.

asked with respect to recrystallization of quartz silt and sand during the process of lithification and thermal metamorphism. At the time of this writing, we have yet to find or develop a technique which can monitor such behavior and express the results in a succinct (and comprehensible) fashion. Two techniques which we consider to be of highest potential are Q-mode factor analysis and Fuzzy K-Means.[16] Our present use of these methods involves determination of similarities between particles within a sample, thus identifying the degree of membership of each particle to each mode. This has been performed in a sequence of samples representing progressively greater and greater recrystallization. A principal problem remaining is the correlation of modes and mode membership from sample to sample, which is under current investigation.

Our experience over the past 10 years has demonstrated that, in terms of sedimentary geology, considerable information of interest to the geological community is carried by particle shape. The easiest part of our task has been the characterization of the shape of a discrete particle. The most difficult task has been to determine appropriate numerical-statistical techniques to extract information from the multivariate data field so generated. Because sustained and concentrated research in this area has not been conducted before, we could not utilize a large store of previously generated data, which forced us to begin our analysis with determination of the nature of the underlying frequency distribution. This, in turn, influenced the subsequent choice of numerical-statistical methods of analysis.

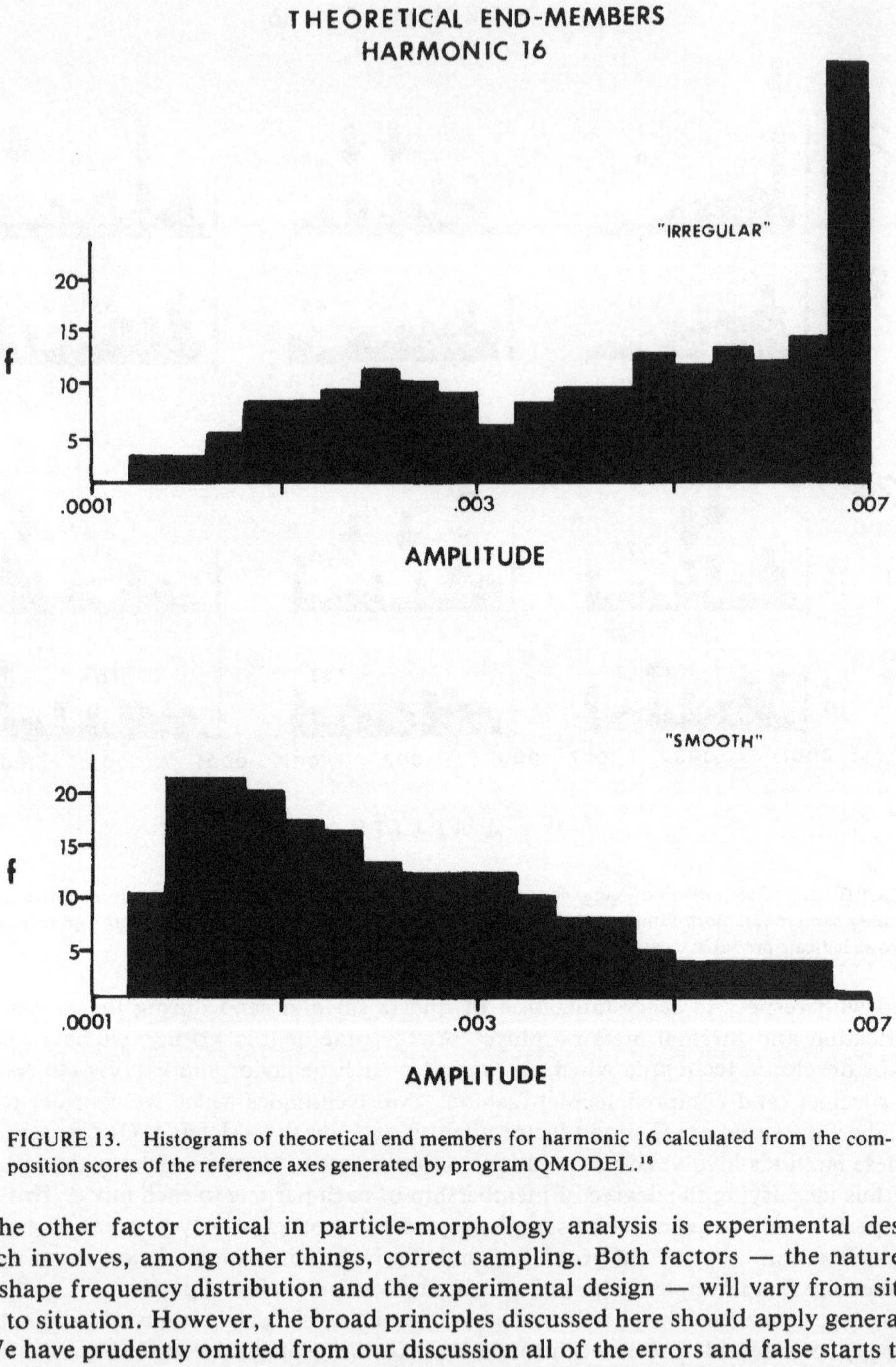

FIGURE 13. Histograms of theoretical end members for harmonic 16 calculated from the composition scores of the reference axes generated by program QMODEL.[18]

The other factor critical in particle-morphology analysis is experimental design which involves, among other things, correct sampling. Both factors — the nature of the shape frequency distribution and the experimental design — will vary from situation to situation. However, the broad principles discussed here should apply generally.

We have prudently omitted from our discussion all of the errors and false starts that have occurred in the course of our work. These have involved all aspects of the discussion above, including bad sampling plans, incorrect use of statistical techniques, and failures in the interpretation of results. These, however, were only identified as errors in retrospect. Because particle-morphology analysis is in its infancy, such errors will probably continue to occur. However, our studies and similar efforts by others have shown that particle shape studies supply unique information on aspects of the natural system difficult or impossible to obtain by other methods.

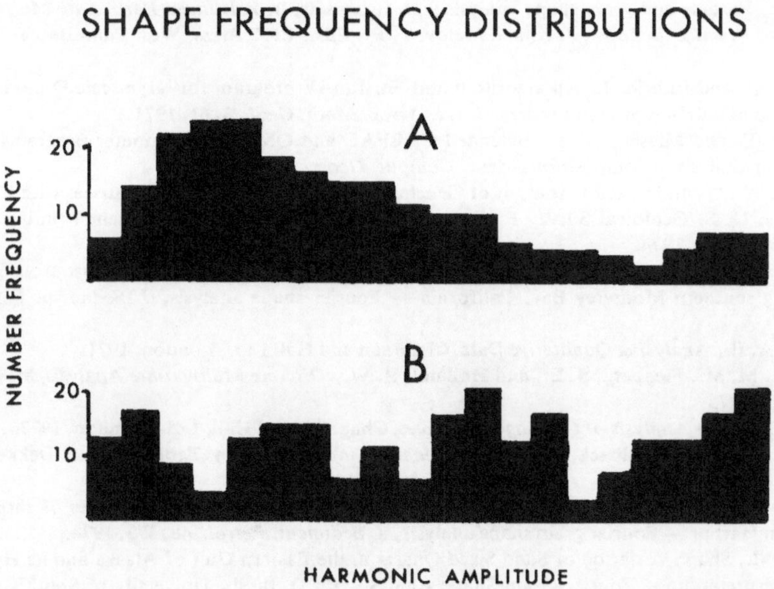

FIGURE 14. Mature (A) and immature (B) shape frequency distributions.

REFERENCES

1. **Tilmann, S. E.**, The effect of grain orientation on Fourier shape analysis, *J. Sediment. Petrol.*, 43, 867, 1973.
2. **Grace, J. T.**, New Microtechnique of Size Analysis for Size-Frequency Characteristics of Sand Laminae, M.Sc. thesis, University of South Carolina, Columbia, 1974.
3. **Redmond, B.**, The Utility of Fourier Estimates of Grain Shape in Sedimentological Studies, M.Sc. thesis, Michigan State University, East Lansing, 1970.
4. **Waltz, S. R.**, Evaluation of Shapes of Quartz Silt Grains as Provenance Indicators in Central Michigan, M.Sc. thesis, Michigan State University, East Lansing, 1970.
5. **Onofryton, J. K.**, The Relationship Between Size and Shape of Sand and Silt Using Fourier Shape Analysis, M.Sc. thesis, Michigan State University, East Lansing, 1973.
6. **Przygocki, R. S.**, Identification of Sediment Sources in a Fluvial Network — North Carolina Blue Ridge — Fourier Grain Shape Analysis, M.Sc. thesis, University of South Carolina, Columbia, 1976.
7. **Pettijohn, F. J.**, *Sedimentary Rocks*, 2nd ed., Harper & Row, New York, 1957.
8. **Ehrlich, R and Weinberg, B.**, An exact method for the characterization of grain shape, *J. Sediment. Petrol.*, 40, 205, 1970.
9. **Beddow, J. K.**, Particle morphological analysis, in *Advanced Particulate Morphology: Theory and Practice*, Beddow, J. K. and Meloy, T. P., Eds., CRC Press, West Palm Beach, Florida, 1979, 3.
10. **Mood, A. M. and Graybill, F. A.**, *Introduction to the Theory of Statistics*, 2nd ed., McGraw-Hill, New York, 1963.
11. **Brunk, H. D.**, *Introduction to Mathematical Statistics*, Ginn and Co., Boston, 1960.
12. **Fisher, R. A.**, *Statistical Methods and Scientific Inference*, 2nd ed., Oliver and Boyd, Edinburgh, 1959.
13. **Jöreskog, K. G., Klovan, J. E., and Reyment, R. A.**, *Methods in Geomathematics 1—Geological Factor Analysis*, Elsevier, New York, 1976.
14. **Davis, J. C.**, *Statistics and Data Analysis in Geology*, John Wiley & Sons, New York, 1973.
15. **Ball, G. H. and Hall, D. J.**, ISODATA, a novel method of data analysis and pattern classification, Stanford Research Institute, Stanford, 1965.

16. **Bezdek, J.**, Particle and grain shape analysis with fuzzy sets, in *Advanced Particulate Morphology: Theory and Practice*, Beddow, J. K. and Meloy, T. P., Eds., CRC Press, West Palm Beach, Florida, 1973,
17. **Klovan, J. E. and Imbrie, J.**, An algorithm and Fortran-IV program for large-scale Q-mode factor analysis and calculation of factor scores, *J. Int. Assoc. Math. Geol.*, 3, 61, 1971.
18. **Klovan, J. E. and Miesch, A. T.**, Extended CABFAC and QMODEL computer programs for Q-mode factor analysis of compositional data, *Comput. Geosci.*, 1, 161.
19. **Miesch, A. T.**, Q-Mode Factor Analysis of Geochemical and Petrologic Data Matrices with Constant Row-Sums, U. S. Geoloical Survey Professional Paper 574-G, U.S. Government Printing Office, Washington, D.C., 1974.
20. **Porter, G.A., Ehrlich, R., Osborne, R. H., and Combellick, R. A.**, Sources and nonsources of beach sand along southern Monterey Bay, California — Fourier shape analysis, *J. Sediment. Petrol.*, in press.
21. **Maxwell, A. E.**, *Analysing Qualitative Data*, Chapman and Hall Ltd., London, 1971.
22. **Bishop, Y. M. M., Fienberg, S. E., and Holland, P. W.**, *Discrete Multivariate Analysis*, MIT Press, Cambridge, 1975.
23. **Everitt, B. S.**, *The Analysis of Contingency Tables*, Chapman and Hall, Ltd., London, 1977.
24. **Gokhale, D. V. and Kullback, S.**, *The Information in Contingency Tables*, Marcel Dekker, New York, 1978.
25. **Van Niewenhuise, D. S., Yarus, J. M., Przygocki, R. S., and Ehrlich, R.**, Sources of shoaling in Charleston Harbor — Fourier grain shape analysis, *J. Sediment. Petrol.*, 48, 373, 1978.
26. **Yarus, J. M.**, Shape Variation of Sand Sized Quartz in the Eastern Gulf of Alaska and its Hydrodynamic Interpretation — Fourier Grain Shape Analysis, Ph.D. thesis, University of South Carolina, Columbia, 1978.
27. **Yarus, J. M.**, Bedrock Identification by Fourier Grain Shape Analysis of Saprolitic Quartz — South Carolina Piedmont, M.Sc. thesis, University of South Carolina, Columbia, 1976.
28. **Swift, D. J. P.**, Continental shelf sedimentation, in *Marine Sediment Transport and Environmental Management*, Stanley, J. S. and Swift, D. J. P., Eds., John Wiley & Sons, New York, 1976. Chap. 15.

Chapter 4

PARTICLE AND GRAIN SHAPE ANALYSIS WITH FUZZY SETS*

James C. Bezdek

TABLE OF CONTENTS

* Partially supported by NSF Grant No. MCS77-00855

ABSTRACT

This chapter relates the Fourier descriptor approach in particulate morphology with certain clustering algorithms based on the axiomatic structure of fuzzy sets. I have abandoned here the direct style currently in vogue in mathematical circles in hopes readers can acquire a real "feeling" for fuzzy sets. The problems of the workshop are cast in my language; then, I discuss the philosophy and characterization of fuzzy sets; following this, I describe the fuzzy ISODATA algorithms, and finally I offer some conjectures on their use in connection with the problems set before us by the workshop.

Because the references in this chapter are in some respects unorthodox, a note of explanation concerning them is indicated. I have used three kinds of references. First, there are formal references which appear in the text and thereafter in the usual fashion. Secondly, a number of extra references which are germane to , but not directly involved in the material of this chapter are collected with remarks in a separate section. Finally, there are occasional allusions to authors and their work which are not formally referenced: these are, without exception, the co-authors of this book, and the work alluded to can be found in other chapters herein. Specifically, I refer to Tom (Meloy), Keith (Beddow), Bob (Erlich), Julius (Tou), Keishi (Gotoh), and Bob (Redinbo).

I. INTRODUCTION

Presuming that a survey of the ground traversed may suggest a good way to choose the path ahead provides for me a way to begin in the context of the workshop. Tom and Keith began by describing the economic significance (in terms of capital, lives, energy, etc.) of several problems arising in particulate technology. It is their conjecture that an efficient and accurate method for the analysis and synthesis of particulate shapes would be useful in addressing some of these problems. Moreover, both felt that the method of Fourier descriptors held great promise for this application. This conjecture was immediately corroborated by Bob Ehrlich, who described an effective method for grain shape analysis based on Fourier descriptors. Variants of this technique were discussed by Keishi; its theoretical underpinnings by Bob Redinbo. The overall effect of these discussions led us to agree that the basic philosophy of particle analysis via Fourier descriptors was theoretically sound and to a certain extent, empirically successful. We conceded, however, that this operational strategy could be improved by implementation of new techniques — both hardware and software — which have evolved recently from the discipline of pattern recognition. Julius suggested that shape description would be improved by extracting more biological features representing Profile and texture. His presentation convinced us that sohpisticated hardware exists which might significantly improve basic methods of feature extraction. Against this backdrop, my role at the workshop became clear: is it possible to augment the method of Fourier descriptors with fuzzy software which might enhance the capability of the existing method. I think the answer is yes. In order to describe my conjectures along these lines, it is necessary to start by defining the problems. From here we pass to the theoretical and philosophical basis of fuzzy sets, and from here we can proceed to a description of the fuzzy ISODATA clustering algorithms. Finally, we can then suggest their role in particulate shape analysis.

II. THE PROBLEMS

Let X = $\{\underline{x}_1, \underline{x}_2, \ldots, \underline{x}_n\}$ R^d. In what follows we regard X an n-sample of points drawn from a physical process or derived from more primordial raw data.Some terms we use frequently are:

R^d	= Feature space
$X' \subset R^d$	= An n sample from R^d
$x_k \in X$	= A feature vector in X
$x_{kj} \in R$	= Measured features of x_k
d	= Number of features per observation

There were, in my estimation, four interrelated problems (P1—4) underlying the goals of the workshop. We can loosely define these as:

P1. Feature extraction — generation of X (from particles).
P2. Feature selection = reduction of $X \rightarrow X' \subset R^d$ where $d' << d$.
P3. Clustering = partitioning of X orX' into subclasses.
P4. Classification = partitioning of $R^{d'}$ into decision regions using X or X' for training.

Although it may be argued that P1 and P2 are parts of the same question, I have always found it useful to separate them conceptually.

Feature extraction begins at the real world interface. For us it meant contouring and digitizing a grain g_k (I would call this the raw data); followed by generation of its Fourier descriptors say $\{(A_i, \alpha_i)_k;\ 1 \leq i < \infty\}$, which I would describe as refined data. Here A_i is the amplitude, α_i the phase angles of the i-th harmonic in the Fourier expansion of the digitized contour. Practical considerations require one to choose from this set of infinite features a subset $\{(A_i, \alpha_i)_k\ 1 \leq i \leq d\}$ (even though we all agreed = 300!). Thus we arrive at a Fourier descriptor data vector $x_k = \{(A_i, \alpha_i)_k\colon 1 \leq i \leq d]$ $\in R^{2d}$ which, it is hoped, contains that information necessary to essentially regenerate the original contour of g_k. A problem of great interest to us was the choice of d: for Ehrlich, d = 20 harmonics seemed quite adequate; for Tom, concern about fine structure suggested d ≈ 2000. Moreover, Julius contended that shape analysis would be improved by extracting raw textural information, e.g., grey level data for the interior of the contour. In any case, the set X can be thought of as being generated by representing n grains in this fashion. It was our consensus that hardware for generation of raw data was quite adequate.

Once X is obtained, one may ask whether the features chosen are sufficient, or redundant, or independent, etc. This begs the question — sufficient for what? If we suspect (or know) that X contains representatives of more than one class, the idea of discriminatory power of the features suggests itself. This I call feature selection: can we select from X certain features (or fuctional combinations thereof) which reduces the dimensionality d of the data while retaining the discriminatory power necessary to achieve clustering or classification?

P2 is clearly dependent on P1, P3, or P4, because the success of a classifier, for example, depends on the features it uses; conversely, for a given set of features, there is undoubtedly an optimal classifier. (The trick, of course, is to match the data and software.) Furthermore, if one knew precisely which features to extract at P1, then P2 would be unnecessary — and so on. We extricated ourselves from this dilemma by agreeing that $\{(A_i, \alpha_i)\}$ were a pretty good set of features for the analysis of particle shape, but conceded that perhaps a rather small d', say d' = 5 or so, might suffice. This type of selection via empirical frequency distributions is currently being studied by Bob Ehrlich, and is an area I feel will respond to the fuzzy approach outlined below.

Clustering in X simply asks for a c-partition of X, $2 \leq c < n$, which exhibits within-data substructure. That is, are there subgoups in X whose members are (mathematically) similar to one another and dissimilar to members of other subclasses? This would occur, for example, if X was a mixture of quartz and feldspar grains. Or more interestingly: do the shapes of all the particles in a homogeneous sample really have a single signature, or are there several signatures corresponding to different modes of formation, etc? One of the great dangers here resides in the fact that similarity as assessed by any mathematical criterion does not necessarily imply similarity at the physical interface: this ambiguity in itself provides a mathematical argument for the use of fuzzy models as hedges. Subdivision of X by clustering was discussed by both Keith and Bob Ehrlich: I believe the fuzzy clustering algorithm described below will provide both with a better strategy than previous techniques. In fact, I propose to solve P2 by solving P3! A more ambitious and exciting possibility lies in discovering not just clusters in X, but functional relationships between the clusters! This is analgous to experimental physics using computers, and I believe we shall find, via fuzzy ISODATA, a way to sometimes accomplish this.

Classification (P4) can be attempted with or without clustering. Keith hopes to devise a comparator which can identify the particle presented to it by shape analysis. This amounts to partitioning feature space with some decision rule, and invariably a labeled sample X is used to train the classifier, i.e., discover the decision rules. I think nearest fuzzy prototypes may provide a start in this direction.

Having the specific problems set before us, we can proceed to devise a plan of attack: it begins here by questioning the rationale of conventional mathematical models.

III. FUZZY SETS

A fundamental method of science is to classify the states of an individual physical process by groups of observations which admit generalization to laws from which we may then try to infer relationships between physical processes. One of the major obstacles in this scheme is the separation of processes into individual observables: in fact, the Gestalt school, the Heisenberg uncertainty principle, and other well established physical paradoxes suggest that this is quite impossible. This establishes the philosophical need for fuzzy sets. There are in real life many classes of nonrandom objects whose boundaries cannot be adequately specified precisely (deterministic models) or with statistical uncertainty (stochastic models).

The need to represent membership of individuals in classes whose boundaries are inherently vague led Zadeh[1] to define a mathematical model which would capture this fuzziness for quantitative analysis. To understand his idea, consider the sets

$$S_1 = \{\text{all people exactly 8 ft tall}\} \tag{1}$$

$$S_2 = \{\text{all people 8 ft} \pm \epsilon \text{ tall}\} \tag{2}$$

$$S_3 = \{\text{all people about 8 ft tall}\} \tag{3}$$

In Equation 2, ε is the tolerance of a measuring device. Sets S_1, S_2, and S_3 are, respectively, conventional (or hard), stochastic, and fuzzy models of the same situation. To discuss their differences, let p denote a person, and h(p) denote his height. The membership of p in these sets is characterized mathematically by values of functions whose common domain is {all people}, but whose ranges differ. The set S_1 can be represented by a characteristic function u_1: {all people} $\rightarrow$ {0, 1} defined by:

$$u_1(p) = \begin{cases} 1; & h(p) = 8 \text{ ft} \\ 0; & \text{otherwise} \end{cases} \quad \text{(1a)}$$

The improbable height of 8 ft notwithstanding, it is easy to see that S_1 is a very sparse hard set. Indeed, when faced with the actuality of determining membership in S_1, it becomes clear that the mathematical exactness expressed by $h(p) = 8$ cannot be physically realized due to errors in measurement. This leads to the incorporation of random error via the tolerance ε. Accordingly, S_2 is replaced by a characteristic function u_2: {all people} → {0, 1} defined by:

$$u_2(p) = \begin{cases} 1; & h(p) = 8 \text{ ft} \pm \epsilon \\ 0; & \text{otherwise} \end{cases} \quad \text{(2a)}$$

Function u_2 is realizable, and from it one can determine the height of p up to statistical uncertainty ε. If $\varepsilon = 0.01$ ft, then $u_2(p) = 1 \leftrightarrow 7.99' \leqslant h(p) \leqslant 8.01'$. Note however, that even though S_2 is a randomized model of this situation, it is still a hard set (i.e., u_2 is a characteristic function)!

Suppose person q has a measured height of 7.97 ft. The information conveyed about q's height by u_2, viz., that $u_2(q) = 0$, is deceptive in the sense that q and p are really quite similar relative to the property (height) which characterizes S_2. This information is captured, however, by fuzzy set S_3. The fuzziness is due to the modifier about in Equation 3, which admits both statistical and nonstatistical uncertainties via a function valued in [0, 1]. For example, we might have u_3: {all people} → [0, 1] defined as

$$u_3(p) = \begin{cases} 1 \; ; & h(p) \in [7.99, 8.01] \\ .99; & h(p) \in [7.98, 7.99) \cap (8.01, 8.02] \\ \vdots & \\ .50; & h(p) \in [5.50, 5.51) \cap (11.50, 11.51] \\ \vdots & \end{cases} \quad \text{(3a)}$$

Values of u_3 exhibit the degree to which height of p agrees with about 8 ft tall. Let's compare the information conveyed about the heights of persons p, q, and r by each of these models using the table below:

$h(p) = 8.00$ ft	$h(q) = 8.01$ ft	$h(r) = 8.02$ ft
$u_1(p) = 1$	$u_1(q) = 0$	$u_1(r) = 0$
$u_2(p) = 1$	$u_2(q) = 1$	$u_2(r) = 0$
$u_3(p) = 1$	$u_3(q) = .99$	$u_3(r) = .98$

A casual inspection of this table reveals that the fuzzy model carries maximal information about the process it represents. The present example would be most convincing for basketball coaches or anthropologists.

In our example above, we call S_3 a fuzzy set because of its verbal description, but in fact its only mathematical realization is via the membership function u_3; the value $u_3(p) \varepsilon [0, 1]$ is called the grade of membership of p in fuzzy set u_3. It may be argued that the value $u_3(p) = .98$ can, by a suitable change in point of view, be interpreted as the probability that $p \varepsilon S_2 = .98$. Such is not the case! To appreciate the philosophical distinction between stochastic and fuzzy models, the workshop performed the following thought experiment: let

$$B = \{\text{all cases of beer}\} \tag{4}$$

Now imagine yourself hopelessly lost in the Smoke Creek Desert, victimized by fate. Having exhausted your meager ration of water, it seems inevitable that you must at last succumb to the vagaries of fate, when at the instant of your demise, two plain cartons, c_1 and c_2, appear ahead. These cartons form with you an equilateral triangle of sides 100 ft and bear only the labels: Prob ($c_1 \varepsilon B$) = .98; Memb ($c_2 \varepsilon B$) = .98, Prob and Memb referring respectively to probability of and fuzzy membership in B. Unfortunately, you have exactly enough strength left to travel 101 ft. Which carton would you decide to open?

Discarding the degenerate case (schizophrenia), and matters of taste as well, members of the workshop chose c_2 without exception. The point is simple: although there is an excellent chance that c_1 is beer, *upon observation* the a posteriori probability for c_1 is Pr($c_1 \varepsilon B$) = 1, if and only if c_1 is beer; and Pr($c_1 \varepsilon B$) = 0 otherwise, in which case c_1 might turn out to be a Tasmanian devil. On the other hand, Memb ($c_2 \varepsilon B$) = .98 must, by any reasonable standards, imply that c_2 is a case of some nontoxic liquid — probably carbonated, perhaps hallucinogenic, etc., and upon observation, the a posteriori membership of $c_2 \varepsilon B$ remains 0.98! In short, fuzzy membership is interpreted as a measure of similarity between features of the individual and characteristics of the class.

Having finished my narrative of how the workshop came to understand the fuzzy set (and had some fun in the process), we can proceed to the formalities necessary to understand how fuzzy algorithms might be brought to bear on the business at hand — shape analysis via Fourier descriptors.

IV. THE ISODATA ALGORITHMS

In what follows, let V_{cn} denote the vector space of all real (c × n) matrices, $2 \leqslant c < n$. If $X = \{\underline{x}_1, \underline{x}_2, \ldots \underline{x}_n\} \subset R^d$, a hard c-partition of X can be characterized by a 0-1 matrix $U \varepsilon V_{cn}$. The elements of $U = [u_{ik}]$ are simply

$$u_{ik} = u_i(\underline{x}_k) = \begin{Bmatrix} 1; & \underline{x}_k \in \text{class } i \\ 0; & \text{otherwise} \end{Bmatrix} ; \quad \begin{matrix} 1 \leq i \leq c \\ 1 \leq k \leq n \end{matrix} \tag{5}$$

In other words, row i of U displays the values of the characteristic function $u_i: X \rightarrow \{0, 1\}$ of the ith partitioning subset (or cluster) in X. For example, the partitioning of X = {1, 2, 40, 91, 92} into the disjoint union {1, 2}∪{40, 91, 92} by two clusters can be represented (up to an arrangement of rows) by the matrix

$$U = \begin{bmatrix} 1 & 1 & 0 & 0 & 0 \\ 0 & 0 & 1 & 1 & 1 \end{bmatrix} \quad (6)$$

The set-theoretic properties of hard c partitions of X translated to matrix terminology yield a subset of V_{cn} which contain all possible c partitionings. We denote this subset by

$$M_c = \{U \in V_{cn} \mid u_{ik} \in \{0, 1\} \,\forall i, k; \; \sum_{i=1}^{c} u_{ik} = 1 \,\forall k; \; \sum_{k=1}^{n} u_{ik} > 0 \,\forall i\} \quad (7)$$

M_c is the set of all nondegenerate (no empty clusters) hard c partitions of X. The number of ways one might attempt to identify an optimal partition of X probably lies in the hundreds (see Tou and Gonzalez[2] or Duda and Hart[3] for excellent discussions of many methods). Of these, a technique originated by Ball and Hall[4] known as hard ISODATA has been employed with success in some quarters. It will be helpful pedagogically to describe this algorithm, in order to understand its fuzzy generalization below. To this end we consider the functional

$$J_1(U, \underline{v}) = \sum_{k=1}^{n} \sum_{i=1}^{c} u_{ik} \,||\underline{x}_k - \underline{v}_i||^2 \quad (8)$$

In Equation 8, $U \in M_c$, $\|\cdot\|$ is a norm on R^d, and $v = (\underline{v}_1, \underline{v}_2, \ldots \underline{v}_c)$ with $V_i \in R^d$, $1 \leqslant i \leqslant c$. J_1 is called the within group sum of squared errors objective functional; its value measures the total squared error incurred by representing all of the $\underline{x}_k$ s in cluster i by prototype $\underline{v}_i$. There is an enormous amount of empirical information about the strategy of choosing as optimal a $\hat{U}$ in M_c so that $(\hat{U}, \underline{v})$ minimizes J_1. Two questions are involved: is J_1 a good criterion of similarity?; and if so, how does one find $\hat{U}$? The first question has both geometric and statistical answers: J_1 performs well when the clusters are essentially ellipsoidal, compact, well-separated, and of roughly equal populations. As any of these varies significantly, J_1 begins to produce bad $\hat{U}$s. The merits of J_1 as a clustering criterion are well argued in the literature (c.f. Ling[5]). Our purpose here is to note that data from a large class of physical processes are amenable to the mathematical similarity assessed by J_1 — the trick is to decide when!

Assuming now that minimizing J_1 is the strategy — how to do it? Many approaches exist; the most popular one is due to Ball and Hall,[4] who suggested a technique of iterative optimization that Dunn[6] later proved was necessary for local minimization of J_1. An informal description follows. (Superscripts here are the number of iterations.)

A. Hard ISODATA (to Minimize J_1)

1. Guess a $U^{(0)} \in M_c$;
2. Calculate the c centroids $\{\underline{v}_i^{(0)}\}$ of the clusters in $U^{(0)}$; that is,

$$\underline{v}_i^{(o)} = \frac{\sum_{k=1}^{n} u_{ik}^{(o)} \underline{x}_k}{\sum_{k=1}^{n} u_{ik}^{(o)}} \quad ; 1 \le i \le c \tag{9a}$$

3. Calculate an updated partition $U^{(1)} \varepsilon M_c$ by clustering each $\underline{X}_k$ to the centroid of minimum distance; that is

$$u_{ik}^{(1)} = \begin{cases} 1; \; \| \underline{x}_k - \underline{v}_i^{(o)} \| = \min \| \underline{x}_k - \underline{v}_j^{(o)} \| \\ 0; \text{ otherwise} \end{cases} \quad \begin{matrix} 1 \le i \le c \\ 1 \le k \le n \end{matrix} \tag{9b}$$

4. Compare $U^{(1)}$ to $U^{(0)}$ by some convenient method. For example, stop if max $\{|u_{ik}^{(1)} - u_{ik}^{(o)}|\} \leqslant \varepsilon$; otherwise return to Equation 9b and update the centroids $\{\underline{V}_i^{(o)}\} \rightarrow \{\underline{V}_i^{(1)}\}$, recalculate $U^{(2)}$ from $\underline{V}^{(1)}$ with Equation 9c, and compare $U^{(2)}$ to $U^{(1)}$.

Of course, this simple scheme of iteration requires care in the matters of initial guesses, tie-breaking rules, convergence criteria, and the like. Aside from the operational details, it makes perfect sense in geometric terms. Keep reallocating the memberships $\{u_{ik}^{(n)}\}$ to clusters with centroids $\{\underline{v}_i^{n-1}\}$ by throwing all the weight $u_{ik}^{(n)}$ for $\underline{x}_k$ on the minimum squared error, i.e., the $\underline{v}_i^{(n-1)}$ closest to $\underline{x}_k$. Moreover, we are in fact minimizing the within-group scatter and maximizing the between-group scatter of the data in the statistical sense of Wilks[3]! (Hence the term minimum variance partitioning for J_1.) Our group found it quite helpful at this point to work through Equation 9 with a numerical example as shown below.

Example A — Let X = {(1, 1), (2, 2), (10, 1), (50, 1), (51, 2)} $\subset R^2$ be 5 points $\{X_1,...X_5\}$ in the plane shown in Figure 1a. Let $\| \cdot \|$ in $\mathbf{R}^2$ be the Euclidean norm, and fix c = 2.

If we choose $U^{(o)} = \begin{bmatrix} 1 & 1 & 0 & 1 & 1 \\ 0 & 0 & 1 & 0 & 0 \end{bmatrix}$, the initial partitioning of X

is $\{\underline{x}_1, \underline{x}_2, \underline{x}_4, \underline{x}_5\} \cup \{\underline{x}_3\}$, as in Fig. (4.1b) (dotted lines). (9a)

2. Now calculate centroids of each cluster using Equation 9a:

$$\underline{v}_1^{(o)} = \frac{\sum_{k=1}^{5} u_{1k}^{(o)} \underline{x}_k}{\sum_{k=1}^{5} u_{1k}^{(o)}} = \frac{(1,1) + (2,2) + (50,1) + (51,2)}{4} = (26, 1.5)$$

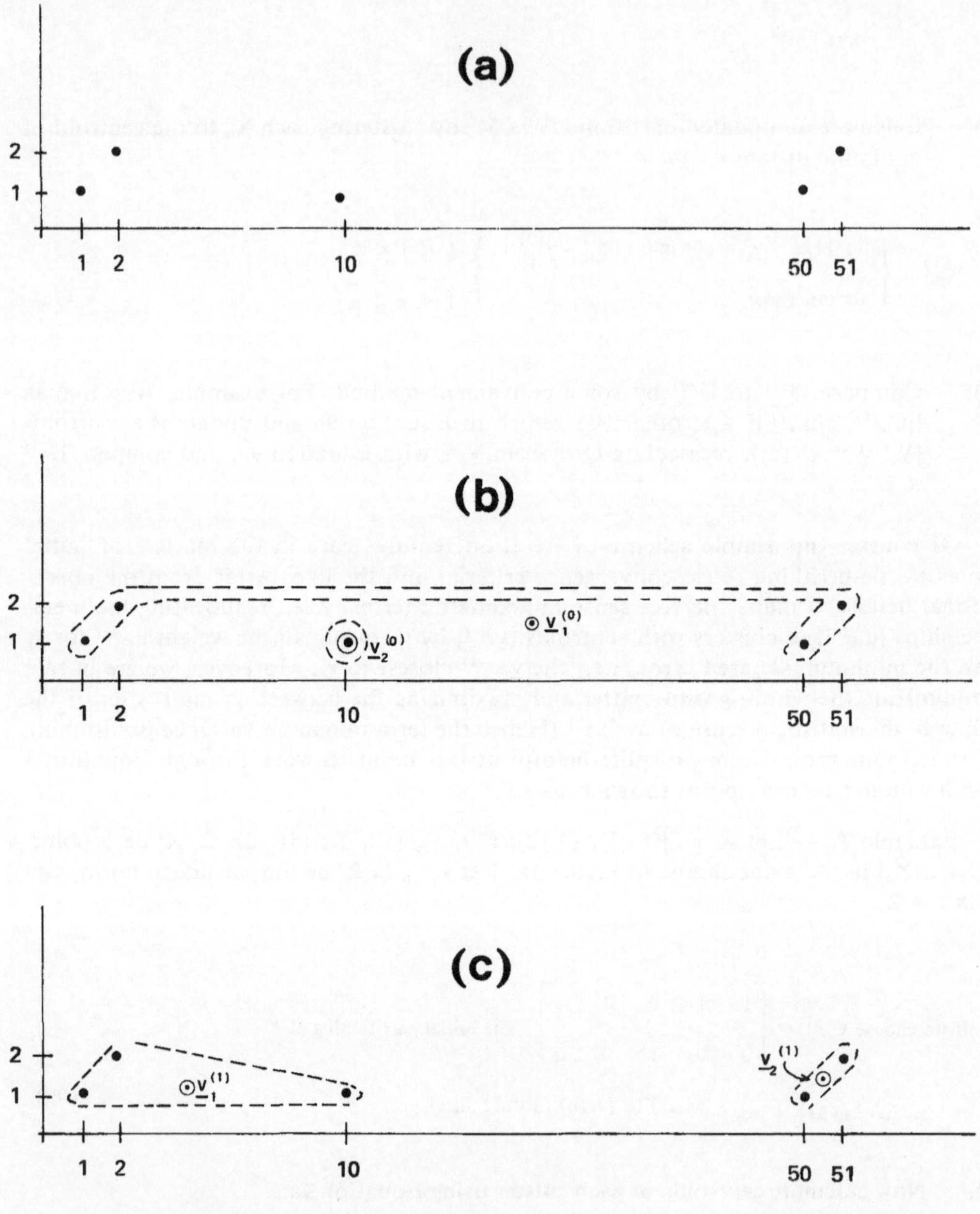

FIGURE 1. Clustering with hard ISODATA (not to scale). (A) The data set X. (B) Initial partioning ($U^{(0)}$, $v^{(0)}$). Dotted lines are hard clusters. (C) One pass with hard ISODATA ($U^{(1)}$, $v^{(1)}$). Dotted lines are hard clusters in $U^{(1)}$.

$$\underline{v}_2^{(o)} = \frac{\sum_{k=1}^{5} u_{2k}^{(o)} \underline{x}_k}{\sum_{k=1}^{5} u_{2k}^{(o)}} = \frac{(10,1)}{1} = (10,1) = \underline{x}_3$$

These centroids appear on Figure 1b which exhibits the pair $(U^{(o)}, \underline{v}^{(o)})$. Now the distance from each $\underline{x}_k$ to $\underline{v}_1^{(o)}$ and $\underline{v}_2^{(o)}$ is calculated, and cluster memberships are reallocated by minimizing distances.

3. For the Euclidean norm, one can see by eye from Figure 1b that the updated partition $U^{(1)}$ via Equation (9b) is

$$U^{(1)} = \begin{bmatrix} 1 & 1 & 1 & 0 & 0 \\ 0 & 0 & 0 & 1 & 1 \end{bmatrix}$$

In other words, $\underline{x}_3$ is regrouped with $\underline{x}_1$ and $\underline{x}_2$, and $\underline{x}_4$, $\underline{x}_5$ become a distinct cluster.

4. Now over all i and k, max $\{|u_{ik}^{(1)} - u_{ik}^{(0)}|\} = 1$. This is deceptive, since three vectors switched allegiance. Nevertheless, J_1 is known to have decreased,[6] and one might decide to stop here. Otherwise, we continue as above, by calculating new centroids:

$$\underline{v}_1^{(1)} = \frac{(1,1) + (2,2) + (10,1)}{3} = (4.33, 1.33)$$

$$\underline{v}_2^{(1)} = \frac{(50,1) + (51,2)}{2} = (50.5, 1.5)$$

The result of the next pass of hard ISODATA appears as $(U^{(1)}, \underline{v}^{(1)})$ in Figure 1c.

As c and n increase, the number of "local" minima of J_1 grows, and this algorithm may stagnate at local trap states (which, incidentally, may not be bad from an inferential point of view). When it is stopped as dictated by Step 4, one has $\hat{U} \in M_c$ with all 0s and 1s. If X is unlabeled, there is very little mathematical analysis at one's disposal to substantiate or repudiate the validity of $\hat{U}$ as a realistic interpretation of substructure in X, a point discussed at length by Ruspini.[7] This problem can be alleviated in the fuzzy case — but more importantly — I strongly object to having to unequivocally classify every point in the data as being unilaterally related to other members of its cluster, and simultaneously unrelated to all other points in X. This is the price one pays for using a hard model; membership reallocation is done on an all-or-nothing basis.

The fuzzy set, on the other hand, would allow each $\underline{x}_k \in X$ to enjoy partial memberships in several fuzzy clusters, as we would anticipate, for example, in any sample representing a mixture of hybrids and progenitors. Thus we are motivated to consider the set M_{fc} M_c defined by

$$M_{fc} = \{U \epsilon V_{cn} : u_{ik} \epsilon [0, 1] \ \forall \ i, k; \ \sum_{i=1}^{c} u_{ik} = 1 \ \forall \ k; \ \sum_{k=1}^{n} u_{ik} > 0 \ \forall \ i\} \tag{10}$$

M_{fc} is the set of all nondegenerate fuzzy c-partitions of X. It was first suggested by Ruspini,[7] and has subsequently been used in the fuzzy clustering alogorithms of Dunn,[6] Bezdek,[8] and Woodbury and Clive.[17] A matrix M_{fc} has no zero rows; columns that sum to one (thus allocating all of the membership of each $x_k \varepsilon X$ over the c fuzzy subsets partitioning X); and rows which exhibit (the value of) c membership functions u_i RX → [0,1], $u_{ik} \doteq u_i(\underline{x}_k)$, which are the fuzzy c partitioning of X. M_{fc} has a number of important mathematical properties which are discussed elsewhere [9-16]; e.g., M_{fc} is compact and is the convex hull of degenerate hard c-partition space. Note that intersections between pairs of fuzzy clusters in U need not be empty, so that subclasses can overlap if the internal strcture of the data so warrants.

With M_{fc} there follows the need for a method to generate optimal fuzzy $\hat{U}$'s: at present the algorithms of Ruspini, Dunn-Bezdek, and Woodbury do this (a study comparing the three outputs on a common data set is under way at present). Our interest focuses here on the Dunn-Bezdek approach. Dunn[6] suggested the generalization of J_1 at Equation 8 to

$$J_2(U, \underline{v}) = \sum_{k=1}^{n} \sum_{i=1}^{c} (u_{ik})^2 \, || \underline{x}_k - \underline{v}_i ||^2 \tag{11}$$

where $U \varepsilon M_{fc}$, $\|\cdot\|$ a norm on $\mathbf{R}^d$, and $\underline{v}$ in $\mathbf{R}^{cd}$, and derived conditions necessary for $(\hat{U}, \underline{v})$ to minimize J_2. Bezdek[8] generalized this to the functional

$$J_m(U, \underline{v}) = \sum_{k=1}^{n} \sum_{i=1}^{c} (u_{ik})^m \, || \underline{x}_k - \underline{v}_i ||^2; \quad 1 \leq m < \infty \tag{12}$$

J_m is a generalized within-group sum of squared errors functional: error $\|\underline{x}_k - \underline{v}_i\|^2$ is weighted now by the mth power of the membership of $\underline{x}_k$ in fuzzy cluster u_i. Under the assumptions that $\underline{x}_k \neq \underline{v}_i \ \forall$ i, k; that $\|\cdot\|$ is differentiable; and that m > 1, it was shown[8] that $(\hat{U}, \hat{\underline{v}})$ could minimize J_m only if

$$\hat{u}_{ik} = \frac{1}{\left(\sum_{j=1}^{c} \frac{|| \underline{x}_k - \hat{\underline{v}}_i ||}{|| \underline{x}_k - \hat{\underline{v}}_j ||} \right)^{\frac{2}{m-1}}}; \quad \begin{matrix} 1 \leq i \leq c \\ 1 \leq k \leq n \end{matrix} \tag{13}$$

$$\hat{\underline{v}}_i = \frac{\sum_{k=1}^{n} (\hat{u}_{ik})^m \underline{x}_k}{\sum_{k=1}^{n} (\hat{u}_{ik})^m}; \quad 1 \leq i \leq c \tag{14}$$

Equations 13 and 14 provide a way to minimize J_m via iterative optimization in exactly the same fashion as hard ISODATA and at m = 1, U is of necessity in M_c; Equation 13 reduces to Equation 9c; Equation 14 to Equation 9e; and $J_m = J_1$. For this reason the algorithms described below are called fuzzy ISODATA:

B. Fuzzy ISODATA (to minimize J_m, $1 < m < \infty$)

1. Guess a $U^{(o)} \varepsilon M_{fc}$;
2. Calculate the c fuzzy centroids $\{\underline{v}_i^{(o)}\}$ of the fuzzy clusters in $U^{(o)}$; that is, calculate initial cluster centers $\{\underline{v}_i^{(o)}\}$ using $U^{(o)}$ with Equation 14.
3. Calculate an updated partition $U^{(1)} \varepsilon M_{fc}$ by reweighting the equivalent total squared errors $\|\underline{x}_k - \underline{v}_i^{(o)}\|^2$ using Equation 13, i.e.,

$$u_{ik}^{(1)} = \frac{1}{\sum_{j=1}^{c} \left(\frac{\|\underline{x}_k - \underline{v}_i^{(o)}\|}{\|\underline{x}_k - \underline{v}_j^{(o)}\|} \right)^{\frac{2}{m-1}}} \; ; \quad \begin{matrix} 1 \leq i \leq c \\ 1 \leq k < n \end{matrix}$$

4. Compare $U^{(1)}$ to $U^{(o)}$ by some convenient method. For example, stop if, over all i and k, max $\{|u_{ik}^{(1)} - u_{ik}^{(o)}|\} \leqslant \varepsilon$; otherwise, return to Step 2 and update the centroids $\{\underline{v}_i^{(o)}\} \rightarrow \{\underline{v}_i^{(1)}\}$ using Equation 14. Then update $U^{(1)} \rightarrow U^{(2)}$ with Equation 13 and compare $U^{(2)}$ with $U^{(1)}$.

This family of algorithms has been studied rather extensively.[8-16] It appears, for example, that m = 2 yields a functional with statistical connotations; that $1.1 \leqslant m \leqslant 1.6$ yields — on the average — best computational results; that the $\{\hat{\underline{v}}_i s\}$ provide a method for feature selection in binary data; and that the $\{\hat{\underline{v}}_i s\}$ can be used to design effective nearest prototype classifiers. To get a feel for the operation of fuzzy ISODATA, we rework Example A with it:

Example B — Let X = {(1,1), (2,2), (10,1), (50,1), (51,2)} C $\mathbf{R}^2$ be the same as in Example A, illustrated again in Figure 2a let $\|\cdot\|$ be the Euclidean norm; and fix c = m = 2.

1. If we choose for $U^{(o)}$ the initial fuzzy 2 partition

$$U^{(o)} = \begin{bmatrix} .9 & .8 & .2 & .7 & .6 \\ .1 & .2 & .8 & .3 & .4 \end{bmatrix}$$

the hard 2 partition closest this $U^{(o)}$ in the sense of maximum membership is the $U^{(o)}$ of Example A.

2. Next, calculate the fuzzy centroids of each cluster:

$$\underline{v}_1^{(o)} = \frac{\sum_{k=1}^{5} (u_{1k}^{(o)})^2 \underline{x}_k}{\sum_{k=1}^{5} (u_{1k}^{(o)})^2} = \frac{.81(1,1) + .64(2,2) + .04(10,1) + .49(50,1) + .36(51,2)}{.81 + .64 + .04 + .49 + .36} = (19.38, 1.43)$$

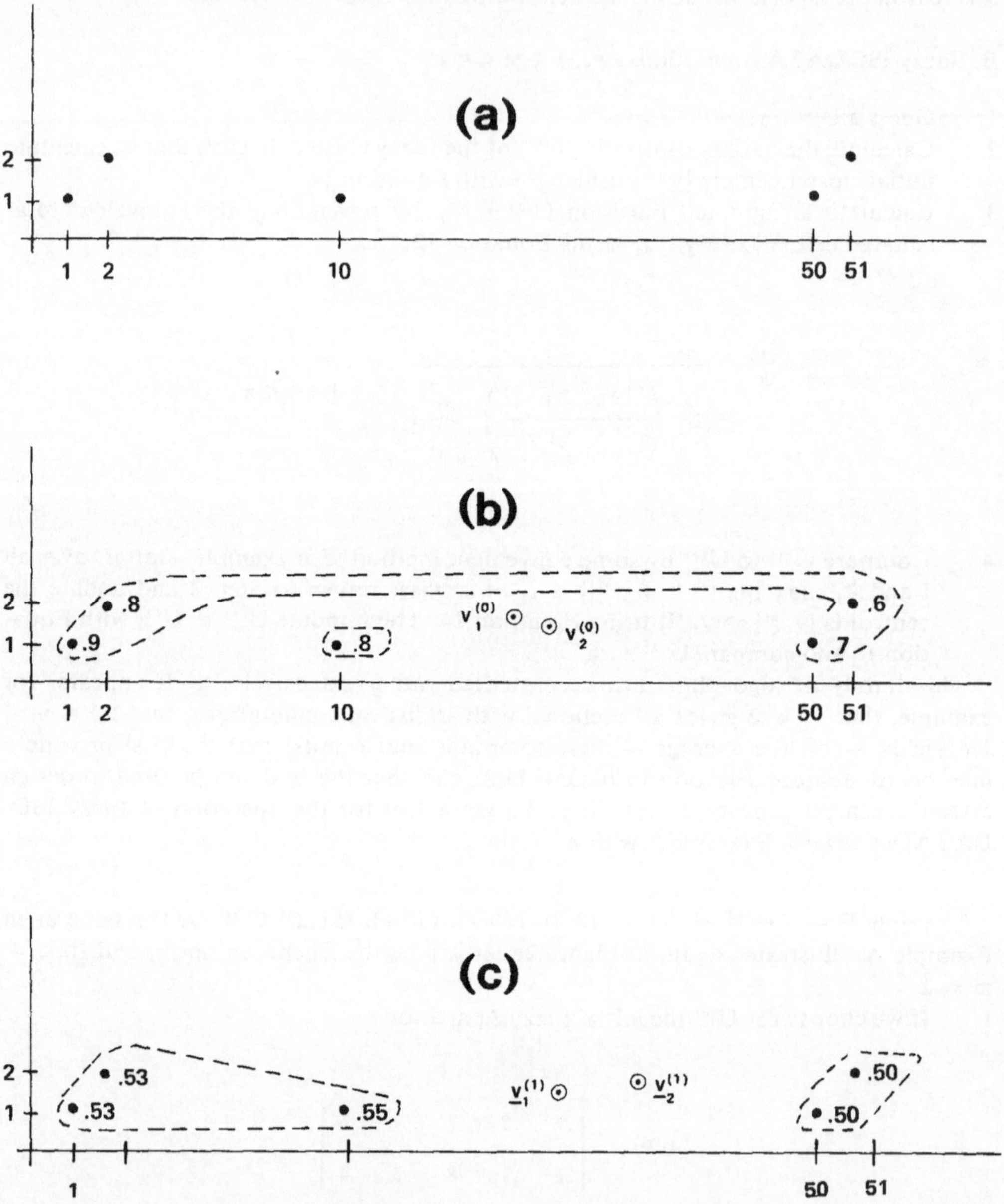

FIGURE 2. Clustering with fuzzy ISODATA (not to scale).(A) The data set X. (B) Initial fuzzy partitioning ($U^{(0)}$, $v^{(0)}$ numbers are maximum memberships in $U^{(0)}$. (C) One pass with fuzzy ISODATA ($U^{(1)}$, $v^{(1)}$) numbers are maximum memberships in $U^{(1)}$.

$$\underline{v}_2^{(o)} = \frac{\sum_{k=1}^{5} (u_{2k}^{(o)})^2 \underline{x}_k}{\sum_{k=1}^{5} (u_{2k}^{(o)})^2} = \frac{.01(1,1) + .04(2,2) + .64(10,1) + .09(50,1) + .16(51,2)}{.01 + .04 + .64 + .09 + .16} = (20.37, 1.21)$$

These cluster centers together with maximum membership values for each $x_k \in X$ appear in Figure 2b, compare them with Figure 1b.

3. Now the 10 distances $\|x_k - v_i^{(o)}\|^2$, $1 \leq i \leq 2$, $1, \leq k \leq 5$ must be calculated, and from them Formula 13 yields an updated fuzzy 2-partition of X. These calculations yield

$$U^{(1)} = \begin{bmatrix} .53 & .53 & .55 & .50 & .50 \\ .47 & .47 & .45 & .50 & .50 \end{bmatrix}$$

Note that x_1, x_2, and x_3 now enjoy maximum membership in cluster 1; x_4 and x_5 are trying to separate from x_1 and x_2 as their memberships in cluster 2 increase. $U^{(1)}$ is fuzzier than $U^{(o)}$; $U^{(2)}$ will start back towards a harder partition as the clusters begin to assert their distinctiveness.

4. Now the maximum membership defect max $\{|u_{ik}^{(1)} - u_{ik}^{(o)}|\}$ = .37. Since $U^{(1)}$ is quite different from $U^{(o)}$, one would probably decide to continue, in which case we find new centroids

$$v_1^{(1)} = (21.39, 1.33)$$

$$v_2^{(1)} = (24.45, 1.41)$$

The results of one pass with fuzzy ISODATA are shown as $(U^{(1)}, v^{(1)})$ on Figure 2c. Note that the fuzzy centroids are starting to migrate towards the apparent visual clusters as the memberships of each x_k are partially reallocated during updating.

Examples A and B will very probably terminate in quite similar partitionings due here to the well-separated structure of X, and in this sense I have not chosen a data set which exhibits the deficiencies of J_1 and the advantages of $J_{m>1}$ (the data was chosen for visual appeal and ease of calculation). As a guess, I would say that hard ISODATA has terminated at $(U^{=1)}, v^{(1)})$; and fuzzy ISODATA would probably terminate with a U something like

$$\hat{U} = \begin{bmatrix} .95 & .97 & .75 & .03 & .04 \\ .05 & .03 & .25 & .97 & .96 \end{bmatrix}$$

which, at the maximum membership level, is the same solution. The important difference is that conversion of $\hat{U}$ to maximum membership is unnecessary, and, in my opinion, less informative than a partition such as in Equation 15, which immediately tells us that x_3 is a bridge of sorts between two pretty separable subclasses. This, of course, is the whole point of fuzzy algorithms! The efficacy of fuzzy ISODATA at ameliorating the difficulties mentioned above in connection with the viability of J_1 as a clustering criterion are well documented.[9-16] Having the mechanics of this algorithm behind us we can proceed to the question: how reliable is a $\hat{U}$ obtained in this way in suggesting substructural relationships in X?

V. CLUSTER VALIDITY

I said above that fuzzy ISODATA would provide a method for assessing the validity of suggested partitionings. To this end consider the functionals on $M_{fc} \to \mathbf{R}$ defined by

$$F_c(U) = \frac{tr(UU^T)}{n} = \frac{\|U\|^2}{n} \quad ; \quad \text{and} \tag{16}$$

$$H_c(U) = \frac{-\sum_{k=1}^{n} \sum_{i=1}^{n} u_{ik} \log u_{ik}}{n} \tag{17}$$

In $\|\cdot\|$ is the Euclidean matrix norm. Partition coefficient F_c is proportional to the average content over all pairwise fuzzy clusters in U; Entropy H_c is a measure of overall fuzzy uncertainty in U. Bezdek[10,11] has derived a number of useful relations for F_c and H_c. In particular, there is for every $U \varepsilon M_{fc}$ and $1 < a < \infty$:

$$0 \leq H_c(U) \leq \log_a c \tag{18a}$$

$$\frac{1}{c} \leq F_c(U) \leq 1 \tag{18b}$$

$$0 = H_c(U) \Longleftrightarrow 1 = F_c(U) \Longleftrightarrow U \epsilon M_c \text{ is hard} \tag{18c}$$

$$\log_a c = H_c(U) \Longleftrightarrow \frac{1}{c} = F_c(U) \Longleftrightarrow U = [\frac{1}{c}] \tag{18d}$$

$$0 \leq 1 - F_c(U) \leq \frac{H_c(U)}{\log_a c} \leq \frac{1}{2}(c - F_c(U)) \tag{18e}$$

The implications of these equations for cluster validity are these: minimizing H_c (or equivalently, maximizing F_c) over the algorithmic parameters $\{\|\cdot\|, U^{(o)}, c\}$ at any fixed m will identify the most separable suggested substructure among the choices considered. In this respect, some comments are in order:

1. This is a heuristic strategy; it has always worked empirically — so far!
2. H_c is a sharper measure of fuzziness than F_c because H_c is essentially logarithmic, F_c parabolic, and H_c has a steeper slope than F_c on (0, 1).
3. $(\log_a c)(1 - F_c)$ is a pretty good lower bound on H_c, and can be used to scale how close H_c is to its minimum value. A normalized bound suggested by Dunn[18] may work even better, namely

$$0 \leq \frac{(\log_a c)(1 - F_c)}{(1 - \frac{c}{n})} \leq \hat{H}_c(U) \doteq \frac{H_c(U)}{(1 - \frac{c}{n})} \tag{19}$$

4. If no choice of $\{\|\cdot\|, U^{(o)}, c\}$ makes H_c approach its lower bound, we do not infer that substructure is absent in X — we infer that fuzzy ISODATA is failing to find any! This is important: the fuzziness here allows the algorithm to be self-flagging (roughly speaking, it tells you when not to trust it).

As an example, we have for $U^{(o)}$ in Example B with logarithms to base e the values

$$F_c(U^{(o)}) = 0.656$$

$$1 - F_c(U^{(o)}) = 0.344$$

$$H_c(U^{(o)}) = .522$$

$$1 - \frac{c}{n} = 1 - \frac{2}{5} = .6$$

so the normalized entropy $H_c(U^{(o)})$ is from Equation 19.

$$0 < 0.573 \leq \hat{H}_c(U^{(o)}) = .870$$

We can use the lower bound as an indication of (absolute) distance, so that $|H_c - \text{lower bound}| = .297$. If fuzzy ISODATA actually arrived at the $\hat{U}$ exhibited in Equation 16, we could compare $\hat{U}^{(o)}$ to U using

$$F_c(\hat{U}) = .867$$

$$1 - F_c(\hat{U}) = .133$$

$$H_c(\hat{U}) = .239$$

$$1 - \frac{c}{n} = .6$$

from which there is the indicator

$$0 < .221 < \hat{H}_c(\hat{U}) = .398$$

So here $|H_c - \text{lower bound}| = 0.177$, which is markedly better. From this it is inferred that fuzzy ISODATA did a pretty good job on X.

There were, at this stage of this author's presentation, quite a few more mathematical details that he found to be interesting, and discussion of feelings about algorithmic protocols, empirical observations, and peculiarities of data in connection with this clustering strategy was desired. Alert as this author was, however, to the fact that equations have a mind-numbing effect on even the most attentive listener, it was discerned that now was the time to relate all of this to the business of the workshop, and leave the burden of these details to references. This decision was met by an enormous sigh of relief! (Although the record will show that the author later reverted to this mode for a speculative excursion into the use of fuzzy relations, which seemed less germane to the immediate interests of the group.) This said, now the payoff.

VI. PARTICLE SHAPE ANALYSIS AND FUZZY ISODATA (OR, WHAT DOES IT ALL MEAN?)

This section contains conjectures for using fuzzy ISODATA in connection with the problems defined in Section B.

P1 — are we extracting the right features — has been addressed in some detail by Julius. Fuzzy ISODATA can tell us little here. For the other problems, however, there is more hope.

P2 — Assuming X to be an aggregate of Fourier descriptor vectors, each of size d, it can be asked whether (1) all d harmonic amplitudes $\{A_i\}$ are significant as descriptors of class shape, and (2) do the d phase angles $\{\alpha_i\}$ contribute significantly to particle shape analysis? These are problems of feature selection, and it was suggested that they can be addressed by clustering columns in X with fuzzy ISODATA. This enterprise is analgous to R-mode factor analysis to reduce dimensionality of the data. Ehrlich and Bezdeck have an investigation of this supposition underway.

P3 — How can clustering with fuzzy ISODATA be otherwise used? Two ways: (1) to assess the actual homogeneity of sets of Fourier descriptor vectors believed to have represented particles of a single type. Keith Beddow and I have begun a study along these lines; (2) to detect functional relationships between subclasses within the data. Ehrlich and I will investigate this possibility very soon.

P4 — How could we use fuzzy ISODATA to design a classifier? Assuming a heterogeneous aggregate of c distinct shape families, it may well be that the cluster centers $\{\hat{v}_k\}$ emanating from clustering this data provides prototypical shape descriptors $\{(\hat{A}_i, \hat{\alpha}_i)_k\}$ for each shape family represented. These signatures could, in turn, be the basis for a nearest fuzzy prototype classifier operating on Fourier descriptor vectors of unlabeled origin.

Whether fuzzy ISODATA will deliver remains to be seen. I want finally to mention that we spent several hours discussing the evolution and status quo of the theory and applications of fuzzy relations. I don't believe that this area is immediately germane to particle shape analysis; on the other hand, I am sure it bears watching for this application. References[19-24] will provide for interested readers a start in the literature in this direction.

VII. BIBLIOGRAPHIC COMMENTS AND REMARKS

Postscripts to scientific papers are — unless they are technical appendices — viewed nowadays with a certain suspicion. However, in view of the unusual and distinctive nature of this workshop, I felt justified in exercising a certain amount of poetic license in this respect. In this last section then, I want to make some observations about the origin of this book, and some remarks of a bibliographic nature which I hope will be useful.

In the first place, I had a lot of fun writing this chapter, which is worth something in itself, and is also an accurate reflection, I believe, of the jocularity enjoyed by us at the workshop. Also at the personal level, the workshop affirmed for me again the conviction that applied mathematics occurs at the interface, not within the interior of the discipline.

More importantly, the atmosphere at this workshop had a very practical effect. I think we extracted from one another vast amounts of information, disparate views, and fresh ideas, all of which can only serve to enrich both individual and collective researches. I'd like to offer a simple explanation: most professional meetings are formal podiums (or battlegrounds) at which much of one's time is spent attending to the usual rivalries, in-house squabbles, and limited points of view. Such was not the case at our workshop, since none of us knew what was "coming next". Because of this, we spent our time learning from and helping one another.

This arrangement could have been a colossal failure had we been unable to communicate effectively. It is my belief, however, that collaborative research arising from our interaction will substantiate that this type of conference is a most effective means of promoting rapid strides in a concentrated research area. To illustrate this point, Reference 25 is a report of some preliminary algorithmic results concerning a generalization of the algorithms discussed in this Chapter. The central idea evolved during our evening work sessions at the workshop: it is an idea of stunning simplicity, but its potential utility in applications quite remote from particulate morphology is enormous.

For those readers interested in the fuzzy sets theory, Zadeh's original paper[1] remains the best introduction that I know of. Several texts on fuzzy sets are available: these include the books of Kauffman[26] and Negoita and Ralescu.[27] An extremely useful bibliography compiled by Gaines and Kohout[28] contains more than 1100 references to papers concerning fuzzy sets that are currently appearing at the rate of 200 or 300 per year. The journal *Fuzzy Sets and Systems* is an excellent source for recent researchers in this area.

More generally, pattern recognition embraces many areas and techniques that are of potential interest and utility for particle and grain shape analysis. I can sidestep my bibliographic duties in this regard simply by mentioning explicitly the excellent texts of Tou and Gonzalez[2] and Duda and Hart[3]: the former excels as an introductory text, the latter as a research monograph and key to the pattern recognition literature. Other books on pattern recognition have been clustered using several criteria by Toussaint.[29]

Ball and Hall[4] suggested the iterative optimization approach for minimization of J_1. Their hard ISODATA algorithm, however, includes a number of embellishments not appearing in Equations 9. The algorithm of Equation 9 is called "Basic ISODATA" in Reference 3, and in the literature it is usually referred to as the k-means algorithm. Accordingly, "fuzzy k-means" would be an historically more appropriate name for fuzzy ISODATA. A sequential version of k-means is discussed in Sebestyn[30] and MacQueen.[31] The work of Ling[5] provides an excellent treatment of the shortcomings of minimum squared error criteria for clustering, as well as an outstanding discussion of graph-theoretic techniques; Zahn[32] is another excellent paper in this area.

Finally, I want to observe that most of the analytical work concerning algorithms for shape analysis appears in journals which may be unfamiliar to readers of this volume. The following is an incomplete, but representative list of publications that often contain papers germane to the questions raised by members of the workshop:

1. IEEE Transactions on Computers
2. IEEE Transactions on Pattern Analysis and Machine Intelligence
3. IEEE Transactions on Information Theory
4. IEEE Transactions on Systems, Man, and Cybernetics

5. Pattern Recognition Journal
6. Journal of the ACM
7. Information Sciences
8. Information and Control

Journals more topical in nature, e.g., *The Journal of Mathematical Biology,* also have papers concerning shape analysis from time to time.

REFERENCES

1. **Zadeh, L.,** Fuzzy sets, *Inf. and Control,* 338, 1965.
2. **Tou, J. and Gonzalez, R.,** *Pattern Recognition Principles,* Addison-Wesley, Reading, Mass., 1974.
3. **Duda, R. and Hart, P.,** *Pattern Classification and Scene Analysis,* Wiley Interscience, New York, 1973.
4. **Ball, G. and Hall, A.,** A clustering technique for summarizing multivariate data, *Behav. Sci.,* 12,153, 1967.
5. **Ling, R.,** Cluster Analysis, Ph.D. Thesis, Yale University, New Haven, 1971.
6. **Dunn, J.,** A fuzzy relative of the ISODATA process and its use in detecting compact, well-separated clusters, *J. Cybern.,* 3, 32, 1974.
7. **Ruspini, E.,** A new approach to clustering, *Inf. and Control,* 15, 22, 1969.
8. **Bezdek, J.,** Fuzzy Mathematics in Pattern Classification, Ph.D. Thesis, Center for Applied Mathematics, Cornell University, Ithaca, N.Y., 1973.
9. **Bezdek, J.,** Numerical taxonomy with fuzzy sets, *J. Math. Biol.,* 1, 57, 1974.
10. **Bezdek, J.,** Cluster validity with fuzzy sets, *J. Cybern.,* 3, 58, 1974.
11. **Bezdek, J.,** Mathematical models for systematics and taxonomy, in *Proceedings of the Eighth International Conference on Numerical Taxonomy,* Estabrook, G., Ed., Freeman, San Francisco, 1974, 143.
12. **Bezdek, J. and Dunn, J.,** Optimal fuzzy partitions: a heuristic for estimating the parameters in a mixture of normal distributions, *IEEE Trans. Comput.,* C-24, 835, 1975.
13. **Bezdek, J.,** A physical interpretation of fuzzy ISODATA, *IEEE Trans. Syst. Man Comput.,* SMC-6, 1976, 387.
14. **Bezdek, J.,** Feature selection for binary data: medical diagnosis with fuzzy sets, in *Proceedings of the National Computer Conference,* Winkler, S., Ed., AFIPS Press, Montvale, N.J., 1976, 1057.
15. **Bezdek, J. and Castelaz, P.,** Feature selection and prototype classification with fuzzy sets, *IEEE Trans. Syst. Man Cybern.,* SMC-7, 87.
16. **Bezdek, J.,** Some current trends in fuzzy clustering, in *Proceedings of the SGSR: The General Systems Paradigm: Science of Change and Change of Science,* Society of General Systems Research, Washington, 1977, 347.
17. **Woodbury, M. and Clive, J.,** Clinical pure types as a fuzzy partition, *J. Cybern.,* 4, 111, 1974.
18. **Dunn, J.,** Indices of Partition Fuzziness and the Detection of Clusters in Large Data Sets, *IFAC World Congress,* Fuzzy Automata and Decision Processes, 1975.
19. **Zadeh, L.,** Similarity relations and fuzzy ordering, *Inf. Sci.,* 3, 177, 1971.
20. **Tamura, S., Higuci, S., and Tanaka, K.,** Pattern classification based on fuzzy relations, *IEEE Trans. Syst. Man Cybern.,* SMC-2, 61, 1971.
21. **Dunn, J.,** A graph theoretic analysis of pattern classification via Tamura's fuzzy relation, *IEEE Trans. Syst. Man Cybern.,* SMC-3, 310, 1974.
22. **Kandel, A. and Yelowitz, L.,** Fuzzy chains, *IEEE Trans. Syst. Man Cybern.,* SMC-4, 472, 1974.
23. **Bezdek, J. and Harris, J.,** Convex decomposition of fuzzy partitions, *J. Math. Anal. and Appl.,* 67 (2), 490, 1979.
24. **Bezdek, J. and Harris, J.,** Fuzzy relations and partitions: an axiomatic basis for clustering, *Int. J. Fuzzy Sets Syst.,* 1, 111, 1978.
25. **Bezdek, J., Gunderson, R., Ehrlich, R., and Meloy, T.,**On the extension of fuzzy k-means algorithms for detection of linear clusters, in Proceedings 1978 IEEE Conference on Decision and Control, 1303, Piscatawy, N.J., January, 1979.

26. **Kauffman, A.,** *Introduction to the Theory of Fuzzy Subsets*, Vol. 6, Academic Press, New York, 1975.
27. **Negoita, C.V. and Ralescu, D.A.,** *Applications of Fuzzy Sets to Systems Analysis*, Birkhauser Verlag, Basle, 1975.
28. **Gaines, B. and Kohout, L.,** The fuzzy decade: a bibliography of books on pattern recognition, *Inf. Sci.*, 8, 1975, 251.
30. **Sebestyn, G.S.,** Pattern Recognition by an Adaptive Process of Sample Set Construction, *IRE Trans. IT*, IT-8, 1962, 82.
31. **MacQueen, J.,** Some methods for classification and analysis of multivariate observations, in *Proc. 5th Berkeley Symp. Math. Stat. and Prob.*, Neyman, J. and LeCam, L.P., Eds., University of Cal. Press, Berkeley, 1967, 281.
32. **Zahn, C. T.,** Graph-theoretical methods for describing and detecting Gestalt clusters, *IEEE Trans. Comp.*, C-20, 1971, 68.

We say that the set is orthonormal if all the constants Γ_n are unity.*

We wish to use this set to linearly approximate certain other functions which are also defined over the same interval I. Our attention will focus upon those functions that possess finite energy, i.e., those f(t), tϵI for which

$$\int_I |f(t)|^2\, dt < +\infty$$

The concept of linear approximation may be stated diagrammatically by

$$\hat{f}_N(t) = \sum_{k=0}^{N} \alpha_k \phi_k(t) \sim f(t) \qquad (3)$$

The constants α_n depend on the functions f(t) and $\{\phi_k(t)\}$ and the approximation is chosen to minimize an error criterion. The approximation error $\varepsilon(t)$ is

$$\epsilon(t) = f(t) - \hat{f}_N(t) \qquad (4)$$

and we employ the mean-square error criterion $\|\varepsilon^2\|$.

$$\|\epsilon^2\| = \int_I |\epsilon(t)|^2 dt \qquad (5)$$

It is very important to understand all of the ramifications of this error criterion, and we will explore some of them in succeeding paragraphs. The constant α_n which minimizes the error criterion appears to depend on $\phi_n(t)$ and f(t) and on all other orthonormal functions $\phi_m(t)$ with m≠n as well as implicitly on the total number of basis functions N.

From the orthonormality and after some manipulations which involve the orthogonality principle,

$$\int_I [f(t) - \hat{f}_{N_0}(t)] \left\{ \sum_{k=0}^{N-1} \beta_k^* \phi_k(t) \right\} dt = 0 \qquad (6)$$

$\{\beta_k\}_{k=0}^{N-1}$ is any set of constants; $f_{n_0}(t)$ is the optimum linear approximation, we may develop the optimum coefficients γ_n which define $f_{NO}(t)$ by

$$\gamma_n = \int_I f(t)\, \phi_n^*(t)\, dt \qquad (7)$$

* There is a well known process by which any linearly independent set of functions can be orthonormalized. It is called Gram-Schmidt orthonormalization.

Also it can be shown that the resulting minimum error is

$$||\epsilon^2||_{MIN} = \int_I |f(t)|^2\, dt - \sum_{k=0}^{N-1} |\gamma_k|^2 \tag{8}$$

The orthonormality properties produce optimum coefficients γ_k which depend only on the respective function $\phi_k(t)$ and the wave-form f(t). The cardinality of the basis set, N, does not enter into the selection of the coefficients; it does impact the approximation error $\|\varepsilon^2\|_{MIN}$ though. The expression for the minimum error, Equation 8 shows that the linear approximation yields an error which is the original energy in f(t), the first term on the right, minus the energy in the approximation. As we increase the number of basis functions, N, the energy in the approximation cannot decrease (the sum of squared magnitudes) and thus the total approximation error $\|\varepsilon^2{}_{MIN}\|$ cannot increase. Therefore we can hope that by increasing the number of basis functions in the orthonormal set, this error can be made vanishingly small. Such is the case for certain countable sets of functions.

In the same vein, we also note that since the error $\|\varepsilon^2\|_{MIN}$ is nonnegative, we obtain the Bessel inequality from Equation 8.

$$\int_I |f(t)|^2\, dt \geq \sum_{k=0}^{N-1} |\gamma_k|^2 \tag{9}$$

Thus if we can extend our basis set to a countable number of functions, i.e., $N \to +\infty$, we still have

$$\int_I |f(t)|^2\, dt \geq \sum_{k=0}^{+\infty} |\gamma_k|^2 \tag{10}^*$$

where we keep in mind that the $\{\gamma_k\}$ are the optimum coefficients selected according to Equation 7.

When the orthonormal basis functions are selected so that equality occurs in the Bessel inequality (10), we say the countable set is complete. This means that the energy in the approximation is identical with the original energy of the signal f(t); the energy in the approximation error ε(t) is zero. Such an equation resulting from this equality is also known as Parseval's relationship

$$\sum_{k=0}^{+\infty} |\gamma_k|^2 = \int_I |f(t)|^2\, dt \tag{11}^*$$

A caveat is in order. The approximation energy may be zero but the actual approximation $\hat{f}(t)$

$$\hat{f}(t) = \sum_{k=0}^{+\infty} \gamma_k\, \phi_k(t) \tag{12}^*$$

may vary wildly from the true wave-form at certain points in I. We reference Papoulis[1] for the dramatic performance for the case of the Fourier basis functions. One must always realize that the approximation is in the mean-square integral sense of Equation 5.

We turn to some concrete examples. The first set of basis functions are indexed by the doubly infinite set of integers.

$$\phi_k(t) = \frac{1}{\sqrt{T}} e^{jk\omega_0 t} \qquad k=0, \pm 1, \pm 2, \ldots$$
$$j=\sqrt{-1},\ I=(0,T) \qquad (13)$$

The fundamental parameter ω_o is related to the interval length T.

$$\omega_0 = \frac{2\pi}{T} \qquad (14)$$

These basis functions are orthonormal as simple integration easily demonstrates, and it can be shown that they also form a complete set. Any function of finite energy defined over the interval (O,T) can be expanded in the mean-square error sense.

$$f(t) = \frac{1}{\sqrt{T}} \sum_{k=-\infty}^{+\infty} c_k\, e^{jk\omega_0 t} \qquad (15)$$

$$c_k = \frac{1}{\sqrt{T}} \int_0^T f(t)\, e^{-jk\omega_0 t}\, dt \qquad (16)$$

The collection of coefficients $\{c_k\}$ is sometimes called the spectrum of f(t). We could treat equally well any function over a general interval by a suitable change of variables. The Fourier series has been used extensively in many branches of engineering, and there is a large body of theoretical and practical results to draw upon.

We will digress for a short exposition on the general Fourier integral transform even though such a transform is for a function defined over an infinite interval and clearly the perimeter of a particle is finite. However, there is an important relationship between this transform and the Fourier series. If we restrict the nonzero region of definition of a function to an interval of length T, we are then investigating the situation which is the topic of this workshop. Furthermore, the concept of convolution (and correlation) which appears in many aspects of signal processing is more easily understood from this viewpoint. The diversion will be brief.

A limiting argument can be applied to the Fourier series to motivate the definition of the Fourier transform. For any f(t) rd Fourier transform is defined over the whole real line with

$$\int_{-\infty}^{+\infty} |f(t)|^2\, dt < +\infty \quad \text{(finite energy)} \qquad (17)$$

the forward Fourier transform is

$$F(\omega) = \frac{1}{\sqrt{2\pi}} \int_{-\infty}^{+\infty} f(t)\, e^{-j\omega t}\, dt\ ; \ \omega \text{ a real number} \tag{18}$$

This may be inverted to recover the original wave-form.

$$f(t) = \frac{1}{\sqrt{2\pi}} \int_{-\infty}^{+\infty} F(\omega)\, e^{j\omega t}\, d\omega \tag{19}$$

An operation which is intrinsic to many physical systems is that of convolution between two wave-forms.

$$c_{xy}(\tau) = \int_{-\infty}^{+\infty} x(\tau - t)\, y^*(t)\, dt \tag{20}$$

The complex conjugate is unnecessary for real wave-forms, but this is the general definition for convolution. In the transform domain, the equation corresponding to convo-lution is

$$C_{xy}(\omega) = X(\omega)\, Y(\omega) \tag{21}$$

where $X(\omega)$, $Y(\omega)$, and $C_{xy}(\omega)$ denote the Fourier transform of the respective wave-forms $x(t)$, $y(t)$, and $c_{xy}(t)$. If the negative sign inside Equation 20 is changed to a positive one, we have the complex correlation between x and y, $r_{xy}(\tau)$. The corresponding transform domain equation becomes

$$R_{xy}(\omega) = X(\omega)\, Y^*(\omega) \tag{22}$$

When we investigate the same operations and their connection with the Fourier series, it is necessary to extend the definition of the functions which we have originally only treated over the interval I. We will denote the periodic extension for a function $f(t)$ defined only for the interval $I = (t_o, t_o + T)$ by $f_p(t)$. Specifically, we may give a defining equation by allowing n to be determined from the stipulation, $nt_o \leqslant t \leqslant nt_o + T$ and

$$f_p(t) = f(t-(n-1)t_0) \tag{23}$$

Then the periodic convolution and correlation functions are given

$$z_{xy}(\tau) = \frac{1}{\sqrt{T}} \int_I x_p(\tau - t)\, y_p^*(t)\, dt \quad \text{(convolution)} \tag{24}$$

$$r_{xy}(\tau) = \frac{1}{\sqrt{T}} \int_I x_p(\tau+t)\, y_p^*(t)\, dt \quad \text{(correlation)} \qquad (25)$$

The respective spectrum $c_{k,z}$ and $c_{k,r}$ for the Fourier series representation of each of these functions is given in the following equations.

$$c_{k,z} = c_{k,x}\, c_{k,y} \quad \text{(convolution)} \qquad (26)$$

$$c_{k,r} = c_{k,x}\, c_{k,y}^* \quad \text{(correlation)} \qquad (27)$$

The second subscript of each coefficient delineates the function to which it applies.

We note from equality (27) the effect of using the squared magnitude of the Fourier series coefficients; it is equivalent to autocorrelation function of the extended version of the original wave-form x(t). Thus using the squared magnitude in particle morphology implies that we are effectively forming the autocorrelation function of the perimeter of a particle. Statistical averaging then may be used to provide an estimate of the average autocorrelation function of a class. (Clearly interval scaling must also be included before any averaging.)

When dealing with the squared magnitude of Fourier series coefficients we wish to make another point. Suppose that we linearly shift the wave-form $x_p(t)$ by a translation of λ

$$x_{ps}(t) \stackrel{\Delta}{=} x_p(t+\lambda) \qquad (28)$$

The spectrum of this shifted signal can be shown to be

$$c_{k,x,s} = \frac{1}{\sqrt{T}} \int_I x_{ps}(t)\, e^{-jk\omega_0 t}\, dt = e^{jk\omega_0 \lambda}\, c_{k,x} \qquad (29)$$

Consequently the squared magnitude of the two spectra are identical. These are sometimes referred to as the power spectrum.

$$|c_{k,x,s}|^2 = |c_{k,x}|^2 \qquad (30)$$

There is a connection between the Fourier series and the Fourier integral transform for a special class of wave-forms — those finite energy signals of finite duration. In particular, consider a finite energy wave-form f(t) where

$$f(t) \equiv 0 \quad |t| > L/2 \qquad (31)$$

The Fourier transform of this signal of duration L is defined by

$$F_L(\omega) = \int_{-L/2}^{L/2} f(t)\, e^{-j\omega t}\, dt \tag{32}$$

On the other hand, if we attempted a Fourier series expansion of the same function we would need the fundamental parameter

$$\omega_0 = \frac{2\pi}{L} \tag{33}$$

With this in mind substitute integral multiples of ω_0 for the transform variable in Equation 32.

$$F_L(k\omega_0) = \int_{-L/2}^{L/2} f(t)\, e^{-k\omega_0 t}\, dt \tag{34}$$

But the right integral is basically the kth Fourier series coefficient of f(t).

$$\frac{F_L(k\omega_0)}{\sqrt{L}} = c_k \tag{35}$$

It is reassuring that for equivalent signals the two transforms can be interrelated; consistency is comforting.

Another example of a complete orthonormal set of functions defined over the interval (0,1) is the Walsh functions. Professor Meloy has successfully applied this set in his work (see Chapters 1 & 2 of this book). By now it should be clear that we may always consider any interval I to be (0,1). For any interval I of the form given by Equation (1) may always be transformed to (0,1) by a suitable linear translation and scaling.

There are numerous labelings for the definition of Walsh functions. We choose to use the one originally due to R.E.A.C. Paley.

$$W(n,t) = (-1)\, S_n(t) \quad n=0,1,2,\ldots \tag{36}$$

The exponent depends on the binary expansions of the index n and the time variable t[6].

$$n = \sum_{i=0}^{-\infty} n_i\, 2^{-i} \qquad n_i = 0 \text{ or } 1, \quad i = 0, -1, -2, \ldots \tag{37}$$

$$t = \sum_{j=1}^{+\infty} t_j\, 2^{-j} \qquad t_j = 0 \text{ or } 1, \quad j = 1, 2, \ldots \tag{38}$$

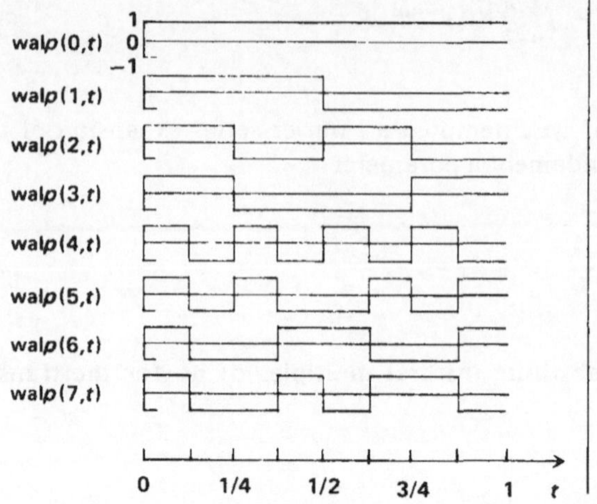

FIGURE 1. First eight Walsh functions.

We always choose a finite expansion for t any time there is an ambiguity (analogous to the two equivalent decimal representations for ½ of 0.5 or 0.49999...; in binary ½ is either 0.1 or 0.0111111111...). Finally we define the exponent for the definition (36) by

$$S_n(t) = \sum_{\substack{i \quad j \\ i+j=1}} \sum n_i t_j = \sum_{i=0}^{-\infty} n_i t_{1-i} \tag{39}$$

The definition of the Paley ordered Walsh functions is finished once we agree to set W(n,t) to zero at any jump points. Clearly W(n,t) can take on either +1 or −1 and must transit between these two values as t varies. There are only a finite number of jump points and the integrals of the type in equation are unaffected by the assignment at them. The first eight Walsh functions are shown in Figure 1. From the general discussion we may explicitly state the Walsh series expansion of a wave-form f(t).

$$\alpha_k = \int_0^1 f(t) \; W(k,t) \; dt$$

$$f(t) = \sum_{n=0}^{+\infty} \alpha_n \; W(n,t) \tag{40}$$

where the reader is reminded that the equality in the latter equation is in the mean-square error sense, i.e., the energy in the corresponding approximation error is zero.

The Walsh functions also generalize to a transform defined for functions over the half-line (0, +∞); the transform variable also is defined over (0, +∞). The binary expansions for the two independent variables have doubly infinite labels now.

$$t = \sum_{j=-\infty}^{+\infty} t_j \, 2^{-j} \qquad t_j = 0 \text{ or } 1$$

$$y = \sum_{i=-\infty}^{+\infty} y_i \, 2^{-i} \qquad y_i = 0 \text{ or } 1 \tag{41}$$

We again stipulate that we use the finite expansion in case of ambiguity. The general exponent for use in the definition of the generalized Walsh functions is $S_y(t)$.

$$S_y(t) = \sum_{\substack{i \; j \\ i+j=1}} \sum y_i t_j = \sum_{i=-\infty}^{+\infty} y_i t_{1-i} \tag{42}$$

$$W(y,t) = (-1)^{S_y(t)} \tag{43}$$

When y assumes integer values this definition is identical with the previous definition for Walsh functions, Equations 36 and 39. Any wave-form f(t) defined for $0 \leqslant t \leqslant +\infty$ can be transformed to a new function F(y), the Walsh integral transform.

$$F(y) = \int_0^{+\infty} f(t) \, W(y,t) \, dt \qquad 0 \leq y < +\infty \tag{44}$$

We need to define the ring sum $\oplus$ on the nonnegative real numbers.

$$r \oplus t = \sum_{i=-\infty}^{+\infty} (r_i \oplus t_i) \, 2^{-i} \tag{45}$$

The binary integers $\{r_i\}$ and $\{t_i\}$ are from the binary expansions of r and t, respectively, and they are combined by the exclusive or operation. (We have used the same symbol for the different but related operations; the context will always distinguish the proper operation.) Since $t \oplus t = 0$, the ring subtraction operation, the inverse of $\oplus$, is identical with the original operation. Therefore all of the previous results concerning the Fourier series and the Fourier transform and the interplay of convolution and correlation with function multiplications in the transform domain are still valid provided all + and − operations in the argument of functions under the integral are replaced by the ring sum operation $\oplus$. (Dyadic correlation and convolution are identical because ring sum and ring subtraction are identical). Consequently when dyadic shifts, i.e., $t \oplus \lambda$, shift of t by λ, are performed in the argument of a function, the magnitude of each of the resulting spectrum terms remains unchanged.

Another complete orthonormal set over the interval (0,1) is the collection of Haar functions. Even though these functions become unbounded as their index increases, they do effectively transform local features of a function into essentially local features. The Haar functions are countable in number, but are indexed by two items

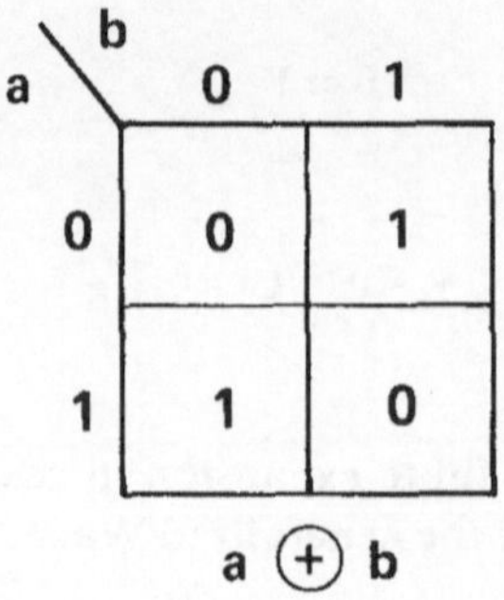

FIGURE 2. Exclusive or operation between binary variables.

$$\mathrm{har}(r,m,t) = \begin{cases} 2^{r/2} & \dfrac{m-1}{2^r} \le t < \dfrac{m-1/2}{2^r} \\ -2^{r/2} & \dfrac{m-1/2}{2^r} \le t < \dfrac{m}{2^r} \\ 0 & \text{elsewhere} \end{cases} \qquad (46)$$

$$r=0,1,\ldots$$
$$m=1,2,\ldots,2^r$$

There is one special function har(0,0,t) which is always one in the interval (0,1). Two features are worth noting. As r increases, the constant nonzero levels in the functions diverge; at index r the functions are nonzero only for a subinterval of length 2^{-r}. Thus as the index increases, the net area of each function remains constant at one. These functions have no known special relationship with convolution or correlation type of operations. They have been successfully used in the design of logic devices.[6]

There is another class of functions which may be useful in examining the morphology of particles. They are not orthonormal but do form a spanning basis for all continuous functions over any closed interval of the real line. We will not formally define them here but the concept of approximation by line segments has been a very useful one.

These basic functions are called piece-wise linear functions (PL functions) and are directly related to the Walsh functions by indefinite integration. The interested reader is referred to the original paper.[7] Several of these functions are displayed in Figure 2. They have a sawtoothed or regular jagged nature which may be quite suitable for the representation of some particles. What is even more appealing is that the proper coefficients for this expansion are derived from a set of linear equations which incorporate regularly spaced samples from the function to be expanded.

III. SAMPLED WAVE-FORMS

So far we have discussed only basis functions for transformations that are defined over a continuum. Signal processing generally involves digital computing machinery which implies the practical consideration of only a sampled version of the particle's perimeter. We insert a brief motivation for the valid sampling of continuous waveforms.

We will consider a signal of finite duration L; specifically the wave-form f(t) obeys

$$f(t) \equiv 0 \qquad t<0\,;t>L \tag{47}$$

A Fourier series expansion requires the fundamental parameter ω_0 and the coefficients c_n.

$$\omega_0 = 2\pi/L \tag{48}$$

$$c_n = \frac{1}{\sqrt{L}} \int_0^L f(t)\ e^{-jn\omega_0 t}\ dt \tag{49}$$

We are now in a position to define the concept of a bandlimited wave-form: f(t) is band limited if

$$c_n \equiv 0 \qquad |n| > N/2 \tag{50}$$

In other words only the first (N + 1) coefficients of the series expansion carry significance. Under these circumstances, f(t) is completely specified by the (N + 1) equispaced samples of the original wave-form.[4]

$$f\left(\frac{kL}{N+1}\right) \qquad k=0,1,2,\ldots,N$$

The exact expression for the f(t) as reconstructed from these (N + 1) samples is

$$f(t) = \sum_{k=0}^{N} f\left(\frac{kL}{N+1}\right) \frac{\sin\left[\left(\frac{N+1}{L}\right)\pi\left(t-\frac{kL}{N+1}\right)\right]}{(N+1)\sin\left[\frac{\pi}{L}\left(t-\frac{kL}{N+1}\right)\right]} \tag{51}$$

What does this require for the equidistance sampling density? There is a minimum of (N + 1) samples necessary over an interval of length L; the spacing between samples, Δt, must satisfy the following inequality for adequate reproduction to be possible.

$$\Delta t \leq \frac{L}{N+1} \tag{52}$$

We may designate the bandwidth of the wave-form f(t) by B.

$$B = \frac{N\omega_0}{2(2\pi)} \qquad (53)$$

The sampling frequency f_s is the reciprocal of the sampling interval, Δt. Inequality (52) can be restated as

$$f_s \geq 2B\left(1 + \frac{1}{N}\right) \qquad (54)$$

Therefore if the sampling density f_s is selected slightly above twice the highest frequency in the input wave-form, sampling does not destroy any of the original information, i.e., we can completely reconstruct the wave-form.

Instead of listing the samples $f(kL/N + 1)$ we can introduce an impulse representation.

$$f_i(t) = \sum_{m=0}^{N-1} \Delta t\, f(m\Delta t)\, \delta(t - m\Delta t) \qquad (55)$$

The function $\delta(x)$ is the Dirac delta function,[1] a generalized function, and N is selected to exceed the sampling rate requirement inequality (54). We choose Δt as

$$\Delta t = L/N \qquad (56)$$

The usefulness of $f_i(t)$ stems from the fact that such a generalized functional representation can be passed theoretically through an ideal filter producing a staircase approximation to the original wave-form $f(t)$.[8] Staircase approximations are easily implemented in practice directly from the samples and are a very common method of wave-form approximation. The area of a step in the staircase approximation is $[\Delta t\, f(m\Delta t)]$, which is the multiplier of the appropriate delta function in the impulse representation $f_i(t)$, Equation 55. We will employ $f_i(t)$ shortly when we discuss the discrete Fourier transform.

IV. DISCRETE ORTHOGONAL TRANSFORMS

One of the most popular transforms in the last decade has been the discrete Fourier transform (DFT). As the name implies, the data are a finite number of samples which we denote as

$$X(m) \qquad m = 0, 1, \ldots, N-1$$

The DFT of these data or data sequence, as it is sometimes called, produces N values.

$$C(k) = \frac{1}{N} \sum_{m=0}^{N-1} X(m)\, W^{km} \qquad (57)$$

$$W = e^{-j2\pi/N} \quad ; \; k=0,1,\ldots,N-1, \quad j=\sqrt{-1} \tag{57}$$

Note the dependence of W on the number of samples N; W is called the Nth root of unity. The inverse DFT bears a close resemblance to the forward transform.

$$X(m) = \sum_{k=0}^{N-1} C(k)\; W^{-km} \tag{58}$$

Convolution and correlation among data sequences again have a corresponding relationship in the transform domain. Hence many filtering and statistical operations may be advantageously performed with transform domain operations and the results inverted to obtain the original required result. Consider the case of convolution.

$$Z(m) = \frac{1}{N} \sum_{k=0}^{N-1} X(\,((m-k))\,)\; Y^{*}(k)$$

$$m=0,1,\ldots,N-1 \qquad ((x)) \text{ denotes modulo N operations} \tag{59}$$

The modulo N additions are equivalent to considering the sequences X(m) and Y(k) to be periodically extended to all integer indices. The corresponding DFT sequences are related as follows where the subscripts denote the appropriate transform sequence.

$$C_Z(k) = C_X(m)\; C_Y(m) \qquad m=0,1,\ldots,N-1 \tag{60}$$

Correlation between the X and Y sequences, again using the modulo N reduction of the arguments, produces R(m) and the following transform domain equations.

$$R(m) = \frac{1}{N} \sum_{k=0}^{N-1} X(\,((m+k))\,)\; Y^{*}(k) \qquad m=0,1,\ldots,N-1 \tag{61}$$

$$C_R(k) = C_X(k)\; C_Y^{*}(k) \qquad k=0,1,\ldots,N-1 \tag{62}$$

The DFT also possesses a shift invariance property. When we shift the data, modulo N, by an amount s, the squared magnitude of the resulting transform coefficients remains unchanged from the ones representing the original data sequence. Let $C_s(k)$ denote the transform coefficients of the shifted sequence X(((m + s))).

$$C_s(k) = \frac{1}{N} \sum_{m=0}^{N-1} X(\,((m+s))\,)\; W^{mk} = C(k)\; W^{-sk}$$

$$k=0,1,2,\ldots,N-1 \tag{63}$$

$$|C_s(k)|^2 = |C(k)|^2 \tag{64}$$

The DFT does have meaning in terms of both the Fourier series and transform. If we employ the impulse representation $f_i(t)$, Equation 55, we easily find its Fourier transform $F_i(\omega)$ is given by

$$F_i(\omega) = \Delta t \sum_{m=0}^{N-1} f(m\Delta t)\; e^{-j\omega m\Delta t} \tag{65}$$

When we narrow considerations to multiples of the fundamental parameter ω_0, Equation 48, we have using Equation 56

$$F_i(k\omega_0) = \frac{L}{N} \sum_{m=0}^{N-1} f(m\Delta t)\; W^{km} \qquad W = e^{-j2\pi/N}$$

$$k=0,1,2,\ldots,N-1 \tag{66}$$

Thus the Fourier transform is simply related to the DFT.

$$F_i(k\omega_0) = L \quad C(k) \tag{67}$$

On the other hand, it can also be shown that the Fourier series of $f_i(t)$ is given by

$$c_k = \sqrt{L} \quad C(k) \qquad k=0,1,\ldots N-1 \tag{68}$$

The complete DFT, i.e., all N terms of the transform sequence, can be computed by a computatially economical method known as the fast Fourier transform (FFT). It is important to realize that the resulting transforms are identical; only the methodology of computing them differs. We will develop the first step in each of two approaches to a fast algorithm. The succeeding steps are just a repetition of the first one. There are other viewpoints and different algorithms.[9]

The first approach we will discuss is called decimation in time. The reason for the name is derived from the fact that we segregate the original data sequence X(m), m = 0,1,..., N-1, into two sequences which are indexed by odd and even integers. We will assume that the number of data points is a power of two; this restriction can be lifted with a little loss in overall efficiency

$$N = 2^n \tag{69}$$

$$Y(r) = X(2r)$$

$$Z(r) = X(2r+1) \qquad r=0,1,\ldots,N/2-1 \tag{70}$$

Then the kth transform coefficient for the original data sequence can be written as

$$C(k) = \frac{1}{N} \left\{ \sum_{r=0}^{N/2-1} X(2r)\ W^{2rk} + \sum_{r=0}^{N/2-1} X(2r+1)\ W^{2rk}\, W^k \right\} \qquad k=0,1,\ldots,N-1 \qquad (71)$$

Identify the first summation by D(k) and the second one, excluding the common factor W^k, by E(k).

$$D(k) = \sum_{r=0}^{N/2-1} Y(r)\ (W^2)^{rk} \qquad (72)$$

$$E(k) = \sum_{r=0}^{N/2-1} Z(r)\ (W^2)^{rk} \qquad (73)$$

We may write for C(k)

$$C(k) = \frac{1}{N}\ \{D(k) + W^k\, E(k)\} \qquad (74)$$

However the term W^2 is a (N/2)th root of unity, exactly the kernel which would be required for a N/2 point DFT.

$$W^2 = e^{-j2\pi/N/2} \qquad (75)$$

A close examination of Equations 72 and 73 reveals that each are a DFT of the shorter data sequences Y(r) and Z(r), respectively.

Note that to compute all N terms of the DFT in a straightforward fashion via Equation 57 would require the following number of operations excluding the scaling by 1/N. (This scaling is just a shift operation of a computer word since N is a power of 2.)

N^2 multiplications

$N(N-1)$ additions

On the other hand, to compute all N terms according to the recipe, Equation 74 requires the following.

$$2\left(\frac{N}{2}\right)^2 + 1 = \frac{N^2}{2} + 1 \quad \text{multiplications}$$

$$2\left(\frac{N}{2}\right)\left(\frac{N}{2} - 1\right) + 1 = \frac{N^2}{2} - N \quad \text{additions}$$

Both numbers have been reduced by almost one-half. Now the same procedure of dividing the sequences in half may be efficiently applied to each of the data sequences of length N/2 involved in the computation of D(k) and E(k). The final result is that the total transform of X(m) can be computed with approximately $N\log_2 N$ additions and $2N\log_2 N$ multiplications.

The second approach to fast transforming that we will briefly explore is called decimation in frequency. Two new sequences are defined; one is the first half of the data sequence while the second one contains the remaining half.

$$U(r) = X(r) \qquad r=0,1,2,\ldots,N/2-1$$

$$V(r) = X(r + N/2) \tag{76}$$

The transform equation may be written as

$$C(k) = \frac{1}{N}\left\{\sum_{r=0}^{N/2-1} X(r)\ W^{rk} + \sum_{r=0}^{N/2-1} X(r + N/2)\ W^{rk}\ W^{N/2k}\right\} \qquad k=0,1,2,\ldots,N-1 \tag{77}$$

The first simplification comes from the identity.

$$W^{N/2} = -1 \tag{78}$$

$$C(k) = \frac{1}{N}\sum_{r=0}^{N/2-1} [U(r) + (-1)^k V(r)]\ W^{rk}$$

$$k=0,1,2,\ldots,N-1 \tag{79}$$

We have not reduced the probem yet to transforms of data sequences with lengths of N/2. However, the frequency variable k may be separated into its odd and even parts, (hence the name of decimation in frequency). For the even frequency terms, we may consider $k = 2p$, and so

$$C(2p) = \frac{1}{N}\sum_{r=0}^{N/2-1} [U(r) + V(r)]\ (W^2)^{pk}$$

$$p=0,1,\ldots,\frac{N}{2}-1 \tag{80}$$

On the other hand, the odd terms are of the form $k = 2p + 1$ and

$$C(2p+1) = \frac{1}{N}\sum_{r=0}^{N/2-1} W^r\ [U(r) - V(r)]\ (W^2)^{pk} \tag{81}$$

These two equations represent the DFT of two data sequences

$$\{U(r) + V(r)\}\ ,\quad \{[U(r) - V(r)]\ W^r\} \qquad r=0,1,2,\ldots,\frac{N}{2} - 1$$

Therefore the odd and even indexed frequency terms of the final transform may be expressed in terms of transforms involving data sequences of length N/2. The number of operations to implement this approach is identical with the previous numbers.

The discrete counterpart to the Walsh series is the discrete Walsh transform (DWT). We will be using the exponential definition of the Walsh functions.[4] The DWT of a finite data sequence X(m), m = 0,1,2,...,N−1, where we are assuming again that N is a power of 2, is defined by

$$B(k) = \frac{1}{N} \sum_{m=1}^{N-1} X(m)\ (-1)^{S_k(m)}$$

$$k=0,1,\ldots,N-1 \qquad N=2^n \tag{82}$$

The exponent in the equation is the inner product of the binary expansions of both k and m.

$$k=k_0 + k_1 2 + k_2 2^2 + \ \ldots + k_{n-1}2^{n-1} \qquad k_i = 0 \text{ or } 1$$

$$m=m_0 + m_1 2 + m_2 2^2 + \ldots + m_{n-1}2^{n-1} \qquad m_i = 0 \text{ or } 1$$

$$S_k(m) = \sum_{i=0}^{n-1} k_i m_i \tag{83}$$

The kernel of the sum in the transform, Equation 82, is either ±1 and so only additions or subtractions are involved. This is an attractive feature for computers that have slow inefficient multiplying capabilities. Clearly, we could develop fast algorithms as before, and there are other fast transform techniques for the class of orthogonal transforms.[4]

The DWT has a shift invariance property with respect to the dyadic operation ⊕. If m and r have binary expansions similar to those of Equation 83, we may define dyadic addition explicitly.

$$m \oplus r = (m_0 \oplus r_0) + (m_1 \oplus r_1)\, 2^1 + \ldots$$

$$+ (m_{n-1} \oplus r_{n-1})2^{n-1} \tag{84}$$

The binary components are combined by the exclusive or operation described in Figure 3. Denoting the transform of the data that have been shifted by r units using the dyadic operation by $B_r(k)$, we have the following relationship.

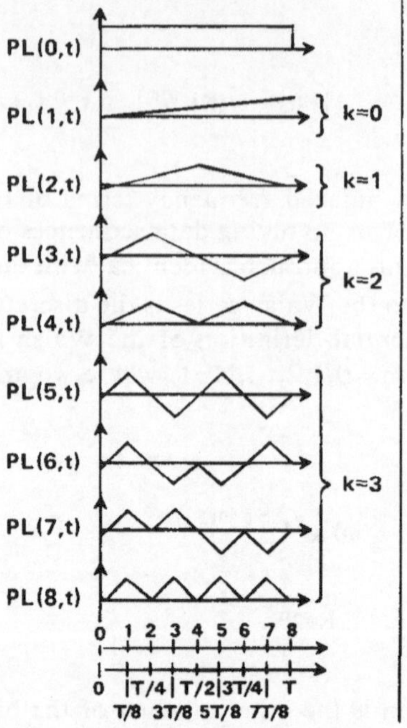

FIGURE 3. Piecewise-linear functions

$$B_r(k) = \frac{1}{N} \sum_{m=0}^{N-1} X(m \oplus r)\ (-1)^{S_k(m)}$$

$$= \frac{1}{N} \sum_{p=0}^{N-1} X(p)\ (-1)^{S_k(p)}\ (-1)^{S_k(r)} = B(k)\ (-1)^{S_k(r)}$$

$$k=0,1,2,\ldots,N-1 \qquad (85)$$

The magnitude of the resulting coefficients are not altered when a dyadic shift is imposed upon the data.

We turn to a relatively recent transform that has shown excellent results when applied to various real signal processing problems, e.g., image data and acoustical waveforms. This class of transforms should be very applicable to particle morphology. The discrete cosine transform (DCT),[10] first proposed in 1974, performs almost as well as the optimum transform which uncorrelates the transform coefficients.[11] (This optimum transform is known as the Karhunen-Loève transform and it will be described shortly at some length.) The data sequence X(m), m = 0,1,...,N−1, is transformed using the set of discrete basis functions.

$$\frac{1}{\sqrt{N}},\ \left\{ \sqrt{\frac{2}{N}}\ \cos\ \frac{(2m+1)k\pi}{2N} \right\} \qquad \begin{matrix} m=1,2,\ldots,N-1 \\ k=1,2,\ldots,N-1 \end{matrix}$$

The transform coefficients for the DCT are labeled by L(k)

$$L(O) = \frac{1}{\sqrt{N}} \sum_{m=0}^{N-1} X(m) \tag{86}$$

$$L(k) = \sqrt{\frac{2}{N}} \sum_{m=0}^{N-1} X(m) \cos \frac{(2m+1)k\pi}{2N} \qquad k=1,2,\ldots,N-1 \tag{87}$$

The inverse transform can be expressed in one generic equation.

$$X(m) = \frac{L(0)}{\sqrt{N}} + \sqrt{\frac{2}{N}} \sum_{k=1}^{N-1} L(k) \cos \frac{(2m+1)k\pi}{2N}$$

$$m=0,1,2,\ldots,N-1 \tag{88}$$

Not only is the reported performance of this class of transforms encouraging, but the transform can be related to the DFT of a modified data sequence. Hence the DCT can be economically mechanized using the FFT. We briefly outline this.[4]

The equation for L(0) remains the same as in Equation 86.

$$L(k) = \sqrt{\frac{2}{N}} \text{ Real part } \left\{ e^{-jk\pi/2N} \left[\sum_{m=0}^{2N-1} \tilde{X}(m)\, W^{km} \right] \right\}$$

$$W = e^{-j2\pi/2N} \tag{89}$$

$$\tilde{X}(m) \begin{cases} X(m) & m=0,1,\ldots,N-1 \\ 0 & m=N,N+1,\ldots,2N-1 \end{cases} \tag{90}$$

The inner bracketed expression is the DFT of the data sequence, length 2N, $\times(m)$, which is X(m) with N zeros appended. The FFT can be invoked for this transform, the extra complex multiplication performed, and the real part extracted to yield the DCT.

V. DISCRETE ORTHOGONAL TRANSFORMS AND STATISTICAL DATA

We investigate transforms from the viewpoint of the statistical properties of the transform coefficients, provided of course that the input data possesses an intrinsic probabilistic description. We will find the use of vector and matrix notation to be helpful. The data is contained in the N vector $\mathbf{X}'$.

$$\mathbf{X}' = \{x_1, x_2, \ldots, x_N\} \qquad ' \text{denotes transpose} \tag{91}$$

The transform is represented by a matrix T and it produces the transform coefficients, the elements of a vector **Y**

$$Y' = \{y_1, y_2, \ldots, y_N\} \quad (92)$$

$$Y = TX \quad (93)$$

The statistical description which we find of importance is the covariance of the vector **X**, Σ_x.

$$\Sigma_X = E\{(X-\bar{X})(X-\bar{X})^{*\prime}\} \quad \text{Covariance} \quad (94)$$

*denotes complex conjugation

$$\bar{X}' = \{E(x_1), E(x_2), \ldots, E(x_N)\} \quad \text{Mean} \quad (95)$$

The covariance of the transform coefficients **Y** can be related to that of the original data vector **X**.

$$\Sigma_y = T\,\Sigma_x\,T^{*\prime} \quad (96)$$

We set two goals for the transform matrix T, both of which fortunately are achievable. The first is that the matrix be a unitary one, or in the case of real entries, an orthogonal matrix. One reason for this choice is the ease with which the inverse transform may be implemented.

$$TT^{*\prime} = I \quad \text{Unitary} \quad (97)$$

I is identity matrix

$$TT' = I \quad \text{Orthogonal} \quad (98)$$

The inverse in each case only involves transposition and, in the former situation, complex conjugation.

The second requirement which we impose is that the covariance matrix of the transform coefficients, Σ_y, be diagonal, i.e., the off-diagonal terms are uncorrelated. The solution yields a matrix for the Karhunen-Lòeve transform (KLT). This transform is sometimes called factor analysis or principal component analysis. The problem is solved using the eigenvectors (characteristic vectors) and eigenvalues (characteristic values) of the data covariance matrix Σ_x. The respective values and vectors are defined by

$$\Sigma_x \phi_i = \lambda_i \phi_i \quad \lambda_i\text{, complex numbers} \quad i=1,2,\ldots,N \quad (99)$$

Then the transform matrix T has these eigenvectors as rows

$$T' = [\phi_1, \phi_2, \ldots, \phi_N] \qquad (100)$$

These rows are linearly independent if all of the eigenvalues are distinct. They also define the covariance matrix, Σ_y, of the transform coefficients.

$$\Sigma_y = T \Sigma_x T^{*'} = \text{Diagonal}\{\lambda_i\} \qquad (101)$$

The Karhunen-Loève transform is dependent upon the exact second-order statistics of the data, and in general, there is no fast transform algorithm for mechanizing Equation 93.

So far we have basically dealt with orthogonal transformations which alter the characteristics of the data to some other domain from which we may recover the original data. This might be termed signal or data representation. But what if we wish not only to transform the data, but also desire to compress this representation? Of course we can no longer invert the transform and recovery of the original data. In order to simplify the demonstration of this data compression problem, we will assume that the data are real valued.

We wish to select an orthogonal matrix T whose rows are vectors $\phi'_1, \phi'_2, \ldots, \phi'_N$ (We use the same notation as above because of the anticipated result.)

$$T' = \{\phi_1, \phi_2, \ldots, \phi_N\} \qquad (102)$$

Remember that the orthogonality of T implies that

$$\phi'_i \phi_j = \delta_{ij} \qquad \delta_{ij} \text{ is Konecker delta function} \qquad (103)$$

But we wish to retain only a subset of the components of the transform vector, **Y**.

$$\mathbf{Y} = T\mathbf{X} \qquad (104)$$

In other words, we retain the components $y_1, y_2, \ldots, y_M$ of the transform elements in Y where M<N. We set the remaining (N−M) positions to some constants b_{M+1}, $b_{M+2}, \ldots, b_N$. The ordering of the components of Y is also part of the problem as is the selection of the constants. Since T is orthogonal, we may reflect the effects of this data compression in terms of the original data **X**. Specifically, for a vector **Y**, the data vector is:

$$\mathbf{X} = y_1\phi_1 + y_2\phi_2 + \ldots + y_N\phi_N \qquad (105)$$

Then the effect in the original data domain of using the M transform coefficients and the (N−M) constants can be expressed.

$$\tilde{X}(M) = y_1\phi_1 + y_2\phi_2 + \cdots$$

$$+ y_M\phi_M + b_{M+1}\phi_{M+1} + \ldots + b_N\phi_N \quad (106)$$

The difference between the vectors of Equations 105 and 106 is of interest.

$$\Delta X = X - \tilde{X}(M) = \sum_{i=M+1}^{N} (y_i - b_i)\ \phi_i \quad (107)$$

Our measure of the loss in the data compression scheme is the mean-square average of ΔX.

$$\epsilon_M^2 = E\{\Delta X'\Delta X\} = \sum_{i=M+1}^{N} E\{(y_i - b_i)^2\} \quad (108)$$

The last equality is a direct consequence of the orthogonality of T. Now we desire to select the rows of T, $\{\phi_i\}$, and the (N−M) constants, $\{b_i\}$, so that ε^2_M is minimized. This problem has a solution via Lagrange multipliers.[4] (There is a constraint among the vectors from Equation 103, $\phi_i'\phi_i = 1$.) The Karhunen-Loêve transform provides the basis for this solution also; the Eigenequation 99 determines the orthogonal matrix T, and the eigenvalues λ_i determine the value of ε^2_m.

$$\epsilon_M^2 = \sum_{i=M+1}^{N} \lambda_i \quad (109)$$

The constants are directly related to the average of the input data by

$$b_i = \phi_i'\bar{X} \qquad i=M+1, M+2, \ldots, N \quad (110)$$

Thus we order the components of Y so that those positions with the least eigenvalues are in the M elements which are discarded.

REFERENCES

1. **Papoulis, A.,** *Signal Analysis,* McGraw-Hill, New York, 1977.
2. **Goldberg, R.R.,** *Methods of Real Analysis,* 2nd ed., John Wiley & Sons, New York, 1976.
3. **Harmuth, H.F.,** *Sequency Theory: Foundations and Applications,* Academic Press, New York, 1977.
4. **Ahmed, N. and Rao, K. R.,** *Orthogonal Transforms for Digital Signal Processing,* Springer-Verlag, New York, 1975.
5. **Redinbo, G. R.,** A note on the construction of generalized Walsh functions, *SIAM J. Math. Anal.,* 2, 166, 1971.

6. **Karpovsky, M. G.,** *Finite Orthogonal Series in the Design of Digital Devices,* John Wiley & Sons, New York, 1976.
7. **Paul, C. R. and Koch, R. W.,**On piecewise-linear basis functions and piecewise-linear signal expansions, *IEEE Trans. Acoust. Speech and Signal Process.*, ASSP-22, 263, 1974.
8. **Taub, H. and Schilling, D. L.,** *Principles of Communications Systems,* McGraw-Hill, New York, 1971.
9. **Brigham, E. O.,** *The Fast Fourier Transform,* Prentice-Hall, Englewood Cliffs, N.J., 1974.
10. **Ahmed, N., Natarajan, T., and Rao, K. R.,** Discrete cosine transform, *IEEE Trans. Comput.*, C-23, 90, 1974.
11. **Shanmugam, K. S.,** Comments on 'discrete cosine transform,' *IEEE Trans. Comput.*, C-24, 759, 1975.

Chapter 6

COMPUTER- BASED PARTICLE SHAPE ANALYSIS FOR CLASSIFICATION, RECOGNITION, UTILIZATION

Julius T. Tou

TABLE OF CONTENTS

I. INTRODUCTION

In materials processing industries, the ingredients as well as some of the products usually exist in three different forms; namely, gaseous form, liquid form, and powder form. Materials in these forms interface with each other during physical transformation, chemical reaction, and mechanical handling. Although most manmade products have been in powder form at one state or another and the bulk tonnage of particles handled by industry is enormous, little basic research work has been done in powder engineering. One of the major reasons that the field of powder dynamics lags behind fluid mechanics and gas dynamics is that the powder problem is not readily amenable to rigorous mathematical treatment which has been successfully applied to gas and fluid problems. The recent advances in image processing and pattern recognition open a new approach to solving the powder problem.

The type and number of particles, the size distribution of each type, and the shapes and texture of the particles may be important measurements upon which decisions may be based. The characterization of particles and powder is quite different from fluid. Particles have been characterized by size, shape, and surface porosity. Historically, such characterization has been accomplished by a variety of sizing techniques, such as sieving, settling, and tapping.[1,2] Particle shape analysis at the present stage of development is similar to particle size analysis some 30 years ago. Shape characteristics of an individual particle may be measured intuitively by distances between parallel tangents of a particle, by comparing the particle with the charateristics of an equivalent sphere, by the length of intercepts, and more recently by Fourier coefficients and by syntactic method. Bulk properties of powders may be measured by tap density, porosity under compaction, flow rate, and permeability. This paper discusses computer analysis of particle shape for classification, recognition, and utilization.

II. BRIEF REVIEW OF IMAGE-PROCESSING AND RECOGNITION

Image processing and recognition[3] has been a subject of intensive study during the past 15 years. It offers a variety of applications, including particle morphology studies. A conceptual block diagram for image processing and recognition is shown in Figure 1. The system consists of four major components: (1) scanning, (2) preprocessing, (3) object isolation, and (4) recognition and interpretation. A TV scanner, a drum scanner, or a flying-spot scanner is commonly used as the scanning system which converts a picture or a scene into a raw digital image. The preprocessing system, which performs smoothing, filtering, enhancement, and formatting, converts a raw digital image into an enhanced and organized digital image of the given picture.

The main function of the object isolation system is to single out the desired objects from the picture for identification and recognition. Object isolation may be accomplished by thresholding, contour finding, edge detection, picture segmentation, and texture analysis. The recognition interpretation system is designed to conduct morphology analysis and scene analysis of the objects so that the image processing system may automatically recognize the objects or the scene and understand the picture. In the case of particle pictures, we are interested in determining the contours of various particles and their textures. The particle contours contain the necessary information for particle identification, and the particle textures provide additional information for recognition and interpretation.

Particles may be characterized by both local and global morphological features. Among the important features are boundary, centroid, area, size, moments, symmetry, elongation index, shape description, major-axis direction, and cluster patterns. These

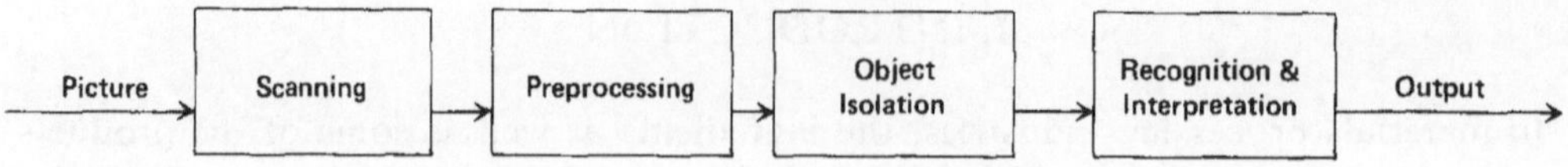

FIGURE 1. An image processing and recognition system.

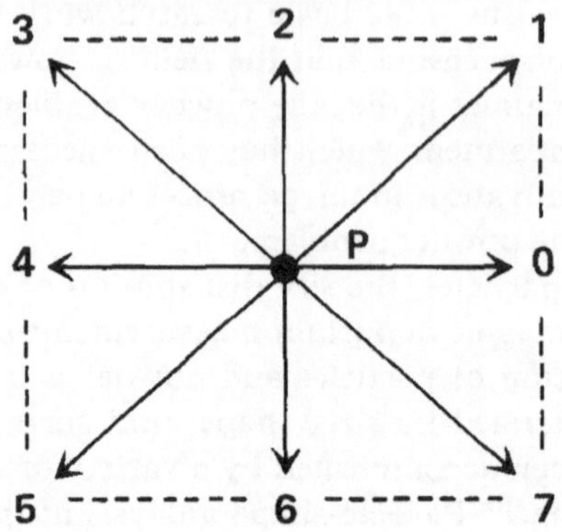

FIGURE 2. Elements of octal chain codes.

parameters, which can be determined by image processing techniques, are useful in performing particle morphology analysis for classification and recognition. To facilitate the analysis of particle morphology by computer, we encode the contours of the particle. The important requirements governing the selection of an encoding technique are: (1) the technique should be efficient in its utilization of machine storage capacity, and (2) the technique should facilitate the operations associated with the pattern analysis procedure. The octal chain encoding technique can satisfy both requirements to a high degree.

The elements of octal chain codes are illustrated in Figure 2. Under this encoding scheme, the location of the next code may assume one of the eight possibilities as shown. These eight codes are labeled 0 through 7. Two standard distances are used, which are in the ratio of $\sqrt{2}$ to each other. One is the side and the other is the diagonal of a unit square in the grid. The octal chain encoding is also referred to as the rectangular-grid encoding. Such an encoding scheme has the following properties. Each point on a curve is located relative to the previous one. Only the first point need be specified in an absolute sense. Only three bits are required to represent one point. Thus, the storage requirement for a continuous curve encoded in this way is only 15% of that required for a curve having all its points independently specified in a 1024×1024 point array. Adding 2 (modulo 8) to each of the digits in the octal code representation of a curve causes the curve to be rotated 90° in the counterclockwise direction. A subtraction will result in clockwise rotation. The length of a curve is given by the number of even digits plus the square root of two times the number of odd digits. Octal chain encoding facilitates the determination of the direction and slope of closed contours by computer. A curve which is representable by octal chain codes in descending order travels in the clockwise direction.A curve which is representable by octal chain codes in ascending order travels in counterclockwise direction.The shape of a particle can be described by the shape of its exterior contour. The second-order differentiation can be used to describe the shape of the particle, since it can partition the particle contour into concave and convex portions.

The problem of pattern recognition and classification may be treated as one of discriminating the input data, not between individual patterns but between pattern classes, via the search for features or attributes among members of a class. In the case of particle studies, particles are recognized and classified according to shape and size, Particles with similar shape characteristics are counted and grouped together. In computer recognition of particle shapes, we may express the measured shape information on a particle in the form of a pattern vector, which represents a point in an n-dimensional measurement space. The set of particles belonging to the same shape class corresponds to an ensemble of points scattered within some region of the measurement space. The particle recognition problem can be viewed as that of generating the decision boundaries which separate the various shape categories on the basis of the measured shape vectors. The decision boundaries may be determined by statistical methods or by syntactic methods. Frequently, the decision boundaries are generated via a training procedure. However, when all of the features of a particle shape category can be determined from sample particles, the shape recognition process may be reduced simply to feature matching.

III. PARTICLE BOUNDARY DETERMINATION

The shape information of a particle is contained in its exterior boundary and interior texture. Boundary contour and texture can be used for the characterization of particle shape. The textural information can be expressed in terms of interior contours and their distribution patterns of the particle. The determination of particle contours is by no means a trivial problem, especially when the picture is noisy. Thresholding is a simple concept which has been commonly used in image-processing studies and boundary determination. When the particles are in sharp contrast with the background, thresholding provides an efficient technique to determine the particle boundaries with reasonable accuracy. To apply the thresholding technique, the picture containing particles is scanned and the intensity histogram is generated. The intensity histogram for the ideal case exhibits two distinct peaks. The boundary threshold may be chosen as the intensity corresponding to the valley between two peaks.[4]

When the particles are not in sharp contrast with the background or the picture is very noisy, the situation gets complicated. The envelop of the intensity histogram is not a smooth curve, and it does not exhibit two distinct peaks. The nonuniform thickness of the particle around the boundary causes fuzziness of the particle contour in the picture. As a result, the valley of the intensity histogram does not represent the boundary threshold accurately. In some situations, the valley may not exist in the histogram. To circumvent these difficulties, Tou has developed a zoom thesholding technique for threshold determination. The zoom thresholding technique determines an optimal boundary threshold through minimization of a performance criterion. The threshold value zooms to the true boundary threshold when the criterion function reaches a minimum.

Once an accurate boundary threshold has been found, we can write a computer program to determine the boundary of each particle in a picture. A simple algorithm for boundary detection is to test the intensity of each picture point against the threshold value. The picture points with intensity equal to the threshold are stored, and other picture points are discarded. Often a smoothing operation is performed on the boundary to fill gaps and to remove noise.

IV. COMPUTER ANALYSIS OF PARTICLE SHAPES

The shape information is essential in performing automatic classification of particles. The particle classification problem may be divided into single-particle classification and aggregate-particle classification. For the former, we analyze each individual particle, and for the latter we study the properties of particle groups. To analyze a single particle, we may consider the boundary characteristics, the textural characteristics, and the three-dimensional characteristics. The boundary characteristics describe the profile information; the textural characteristics contain the surface information including roughness, patterns, and porosity; and the three-dimensional characteristics provide the volumic information. To study aggregate particles, we may consider contour patterns, textural characteristics, and three-dimensional characteristics. These characteristics describe the global features of the particle aggregate.

The shape information of a particle can be derived from its boundary by several techniques: (1) orthogonal function/series expansion, (2) structural analysis, (3) morphological transformation, and (4) stereological analysis. Earlier attempts were made to characterize shape information on particles by orthogonal functions or series. The coefficients of an orthogonal expansion of measurements taken from the particle boundaries are used to describe the particle shape. The orthogonal functions commonly used are Fourier series, Walsh function, Haar function, and z-transforms. This approach has been extensively discussed in the literature[5-7] and will not be elaborated further.

The orthogonal-function approach involves numerical computations. An alternate approach makes use of the structural properties or syntactic relationships which characterize the particle boundaries. Structural characterization of particle boundaries may be accomplished by chain coding, by picture grammar, or by property concatenation. A particle boundary can be described by a heirarchical structure of primitive elements analogous to the syntactic structure of languages. This approach permits application of formal language theory to the particle classification problem. A picture grammar is considered as consisting of finite sets of elements called variables, primitives, and productions. The rules of production determine the types of grammar. The essence of this approach to particle shape classification lies in the selection of boundary primitives, the assembling of the primitives and their relationships into picture grammars, and analysis and recognition in terms of these grammars.

The morphological transformation approach[8] offers a technique for automatic classification of aggregate particles. A morphological transformation which is applicable to certain types of particles is the elliptical transformation. In this approach, the two-dimensional digital image of a particle picture is transformed into a one-dimensional pattern vector characterizing the aggregation of particle orientation, distribution, and shape. Such a pattern vector may be expressed in terms of a vector of quantized slopes from the equivalent ellipse which characterizes the particle shape. These vectors are referred to as the primitives or basis vectors for the aggregate particle picture. We then perform an elliptical transformation on the particle picture. The pattern vector characterizing the combined slope information of all the particles in the picture is decomposed in terms of these basis vectors. Such a decomposition provides quantitative morphological information for automatic classification of particles.

In addition to boundary properties, particle shape is also characterized by textural properties. The shape information may be determined from textural characteristics by the two-dimensional time series approach, the gradient-distribution-matrix approach, or the morphological transformation approach.[8-11] Since particles are three-dimensional objects, the classification of particle shapes may require volumic information which is contained in three-dimensional characteristics. Computer techniques for the determination of three-dimensional characteristics include (1) stereological analysis, (2) iso-intensity contour analysis, and (3) sectional superposition approach.

V. CONCLUSIONS

This paper introduces image processing and pattern recognition approach to particle shape analysis by computer. Some aspects of image processing and pattern recognition are reviewed as a prelude to computer analysis of particle shape. Several techniques for automatic determination of particle boundaries are discussed. The shape information of a particle is contained in its exterior boundary and interior texture. From the boundary, the shape information may be derived by orthogonal function/series expansion, by structural analysis, or by morphological transformation. From the texture, the shape information may be derived by two-dimensional time series analysis, by gradient-distribution-matrix analysis, or by morphological transformation analysis. The texture also contains three-dimensional information on the particle. It is the author's hope that this paper will be able to stimulate more interest in applying image processing and pattern recognition to computer analysis and automatic classification of particles in powder engineering.

REFERENCES

1. **Davis, R.,** NSF Workshop on Particle Technology, Philadelphia, Pa., August 1975.
2. **Beddow, J. K., Vetter, A. S., and Sisson, K.,** Powder metallurgy review 9, particle shape analysis, *Powder Metall. Int.*, 8 (2), 1976.
3. **Tou, J. T and Gonzalez, R. C.,** *Pattern Recognition Principles,* Addison-Wesley, Reading, Mass., 1974.
4. **Tou, J. T , et al.,** Computer analysis of stereological parameters from lung tissue, *Proc. IEEE 1978 Conf. on Pattern Recognition and Image Processing,* Chicago, Ill., 1978.

4a. **Tou, J. T.,** Zoom thresholding technique for boundary determination, *Int. J. Computer Info. Sci.,* 8(1), 1979.

5. **Schwarcz, H. P. and Shane, K. C.,** Measurement of particle shape by Fourier analysis, *Sedimentology,* 13, 1969.
6. **Harmuth, F.,** *Transmission of Information by Orthogonal Functions,* Springer-Verlag, Berlin, New York, 1970.
7. **Tou, J. T.,** *Digital and Sampled-Data control systems.* McGraw-Hill, New York, 1959.
8. **Tou, J. T.,** Automatic analysis and interpretation of cell micrographs, in *Digital Processing of Biomedical Images,* Preston, K. and Onoe, M., Eds., Plenum Press, New York, 1976
9. **Tou, J. T., et al.,** Pictorial Texture Analysis and Synthesis. *Proc. 3rd Int. Joint Conf. on Pattern Recognition,* Coronado, Calif., 1976.
10. **Tou, J. T., et al.,** An approach to Texture Pattern Analysis and Recognition *Proc. IEEE 1976 Conf. on Decision and Control,* Clearwater, Fla., 1976.
11. **Tou, J. T., et al.,** Picture Understanding by Machine via Textural Feature Extraction. *Proc. IEEE 1977 Conf. on Pattern Recognition and Image Processing,* Troy, N. Y., 1977.

Chapter 7

LIQUID AND POWDER IN VIEW OF RANDOM PACKING STRUCTURE — MORPHOLOGY OF PARTICLE ASSEMBLAGES

Keishi Gotoh

TABLE OF CONTENTS

I. INTRODUCTION

Although structures of molecular configurations change at every moment due to thermal vibration, liquids can be regarded as random packing of particles. Since Bernal[3,5] and Scott[21,22] first proposed this idea, many investigations have been made into the spatial distribution of particles. The structural study can play important roles in many fields of science and engineering. The radial distribution function in statistical thermodynamics can express physicochemical properties of matter. Fluidization in chemical engineering is a structural problem. If we take physical and chemical properties away from our subject, the remainder is the structure itself. Hence, the structural study is ubiquitous and of fundamental importance.

Focusing attention on the structural study, this text summarizes a series of the author's research works on the comparison between packing structures of molecular systems and powder assemblages. In what follows, we first discuss the structure of random assemblage of equal spheres. We then discuss the packing situation of molecular systems through a simplified statistical thermodynamics. And, finally, we compare the packing structures between molecular systems and powder assemblages.

It is better for us to explain final results first. Figure 1 illustrates a general comparison between packing structures of molecular systems and powder assemblages, which compares the two systems in terms of the bulk-mean volume fraction ϕ of the particles; λ is the molecular shape parameter. For a while, we will consider the case of $\lambda = 1$ only, i. e., the case of spherical particles. All numerical values at boundaries of the regions are obtained experimentally both from the sphere assembly and from the thermodynamics of simple fluids. First consider the molecular system. When the concentration of molecules is very low, each molecule behaves as an individual and is in the gaseous state. At the critical point, the average spacing of the neighboring particles becomes only about 10% of the particle diameter, and the molecules are considered to be in quasi-contact due to attractive forces among them. Accordingly, in conentrations higher than the critical point, the molecules behave as the gas-like liquid. The region between the boiling and melting points corresponds to what we call the liquid state, and the molecules are in mutual contact. The boiling point corresponds to the random loose packing of particles; the melting point to the random close packing. In the concentration higher than the melting point, the random packing cannot exist. As a result some regularity should appear. This means the occurrence of crystalline. In summary, Figure 1 shows a generalized classification of the state of matter based upon the bulk-mean volume fraction of the particles. From random packing of equal spheres we have experimental evidences that the bulk-mean volume fraction of particles is 0.6366 for the random close packing; 0.601 for the random loose packing. Computer-simulated loose packing[25,26] yields $\phi = 0.582$, while the spherical powder system gently settled down after fluidization[17] yields $\phi = 0.56$. The closest packing density of equal spheres is of course 0.74. Although there is a rough explanation,[11,15] no mathematical procedure can explain all of these critical packing densities. We do not know how to express the packing situation in detail. In the concentration lower than $\phi = 0.55$, each particle is not in contact with others overall, so we can describe this region as being the particle dispersion. As mentioned before, in the range higher than $\phi = 0.235$, the particles are in quasi-contact due to attractive forces among them. The range between $\phi = 0.55$ and 0.64 corresponds to the random packing. The range higher than $\phi = 0.64$ should correspond to the regular packing. However, there are only two types of the regular packing in this region, namely the tetragonal-sphenoidal and rhombohedral packings. Accordingly, if a continuous change in the packing density does occur, a randomness other than the previous criterion must appear in the system, such as the

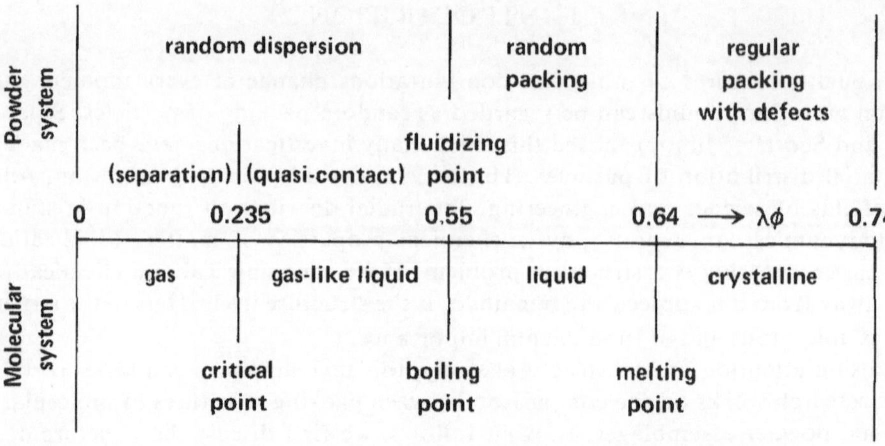

FIGURE 1. General comparison between packing structures of molecular systems and powder assemblages. λ = molecular shape parameter ($\lambda = 1$ for spheres). ϕ= bulk-mean particle volume fraction. (From Gotoh, K., *Nature Phys. Sci.*, 232,1971, 65. With permission.)

regular packing with randomly distributed defects. In this fashion we have obtained the general comparison between packing structures of molecular systems and powder assemblages as shown in Figure 1. As mentioned before, no mathematical procedure is available for explaining the detailed structure of the particle assemblage.

II. STRUCTURE OF RANDOM PARTICLE ASSEMBLAGE

Now we proceed to more detail about the structure of the random particle assemblage. At the packing density of $\phi = 0.55$ we divide the system into two regions: random dispersion and random packing. This boundary concentration is not a rigorous value.

A. Random Dispersion of Equal Spheres

The radial distribution function,[18,24] i.e., the probability of finding a particle at a distance from a central sphere, and the distribution of the local particle concentrations[9] are useful for expressing the structure of the random particle assemblage. However, in the case of low concentration, we can discuss the internal structure in more detail by introducing the distribution function of the particle spacings between the two nearest neighbors. Figure 2 shows the relation between the bulk-mean particle volume fraction and the average nearest-neighbor spacing. The computer experiment is conducted as follows. Three random numbers generated by a computer built-in function determine a position of a sphere subject to the condition that it never overlaps with all others. In this manner we constructed a random dispersion of 10,000 equal spheres, from which we obtained the average spacing of two neighboring spheres. As one can see from Figure 2 in the concentration higher than $\phi = 0.2$, the average spacing becomes less than about 10% of the particle diameter. Accordingly, the particles are in quasi-contact due to attractive forces among them.

In order to explain this relation theoretically, we follow the method of Bansal and Ardell.[1] Let $f(\mathrm{R})\mathrm{dR}$ denote the probability that the center-to-center distance between a particle and its nearest neighbor lies between R and $\mathrm{R} + \mathrm{dR}$ from the given central particle, where R is in units of the sphere diameter. Following Chandrasekhar,[7] $f(\mathrm{R})\mathrm{dR}$ can be expressed as the product of two probabilities: (1) the probability that

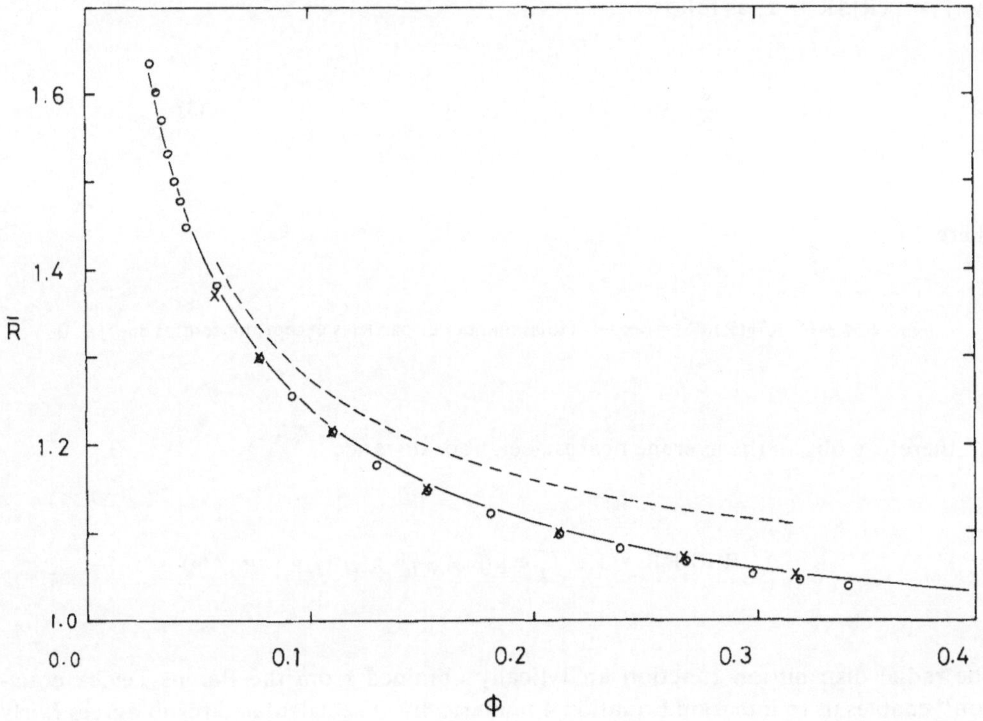

FIGURE 2. **Relation between average near-neighbor spacing R'''s and bulk-mean particle volume fraction ϕ.• = Computer expt.; ———— = theoretical, Equation 4; and — — — — = Bansell-Ardell[1] Solution. (From Gotoh, K., Jodrey, W. S., and Tory, E. M., *Powder Technol.*, 21, 1978, 286. With permission from Elsevier Sequoia, Lausanne.)**

no other particle center exists within the spherical volume of radius R; (2) the probability that a particle center is contained within the spherical shell of radius R and thickness dR. We thus obtain Equation 1:

$$f(R)dR = \left[1 - \int_1^R f(R)dR\right] \cdot f_o(R)dR \qquad (1)$$

The second term in the right-hand side of Equation 1 can be expressed by:

$$f_0(R)dR = 4\pi R^2 dR \cdot N_\nu g(R) \qquad (2)$$

where $N_\nu = 6\phi/\pi$, N_V is the average number density of the particles, ϕ is the bulk-mean particle volume fraction, and g(R) is the radial distribution function. After substituting Equation 2 into Equation 1, we differentiate it with respect to R to obtain a differential equation for f(R), which can be solved with the condition for the probabil-

ity, $\int_1^\infty f(R)dR = 1$, as follows.

$$f(R) = -\frac{d}{dR}\exp\left[-24\phi\int_1^R R^2 g(R)dR\right] \qquad (3)$$

where

$$\exp\{-24\phi\int_1^\infty R^2 g(R)dR\} = \exp\{-(\text{total number of particles except the central one})\} = 0$$

We therefore obtain the average nearest-neighbor distance.

$$\bar{R} = \int_1^\infty R\cdot f(R)dR = 1 + \int_1^\infty \exp\{-24\phi\int_1^R R^2 g(R)dR\}dR \qquad (4)$$

The radial distribution function analytically obtained from the Percus-Tevick equation[24] enables us to integrate Equation 4 numerically. The calculated result agrees fairly well with the computer experiment as shown in Figure 2.

B. Random Packing of Equal Spheres

Since each particle is in contact with others overall, the particle spacing no longer makes sense. The radial distribution function is the most suitable expression for the average internal structure of the packing, and various theoretical approaches to obtain this function have been devised in statistical thermodynamics, but no procedure is expected to apply for the high concentration region such as the random packing. In order to examine the internal structure of the packing, the number of neighboring particles about a central particle can be a good measure. Table 1 is a result from about 8,000 random close packings of ¼-inch bearing balls.[8] As one can see from this table, the cumulative number of neighbors has a wide frequency distribution, suggesting that the packing can be expressed by a mixture model of various aggregates. Owing to the inevitable experimental error,[6] we cannot elucidate how the distribution becomes in the limiting state of the contact ($R \rightarrow 1.0$). From the average number of neighbors listed at the right column in Table 1 we estimate the average contact number of 6.0. The random loose packing of Tory et al.,[25] i.e., a computer simulation of the slow settling of equal spheres into a randomly packed bed yields also the average contact number of 6.0. Is this number an integer and independent of the packing density? We do not have a rigorous answer. A rough explanation might be as follows: for a given sphere to be stable against displacement in a given direction, it must be supported by three others in that direction. Thus taking the opposite direction into account, the contact number of the spheres becomes six.[2,15]

TABLE 1

Frequency Distribution of Cumulative Number of Neighboring Spheres in Random Close Packing of Finney

R(−)	Cumulative number of neighboring spheres															
	1	2	3	4	5	6	7	8	9	10	11	12	13	14	Total	Average
1.005	0	4	25	54	117	164	121	55	15	1	0	0	0	0	556	5.935
1.01	0	4	9	28	109	201	169	143	34	4	0	0	0	0	701	6.521
1.09	0	0	0	0	0	8	57	127	188	122	25	2	0	0	529	8.835
1.20	0	0	0	0	0	0	2	15	70	164	193	64	6	0	514	10.453

From Finney, J. L., *Proc. R. Soc. (London)*, A319, 479, 1970. With permission.

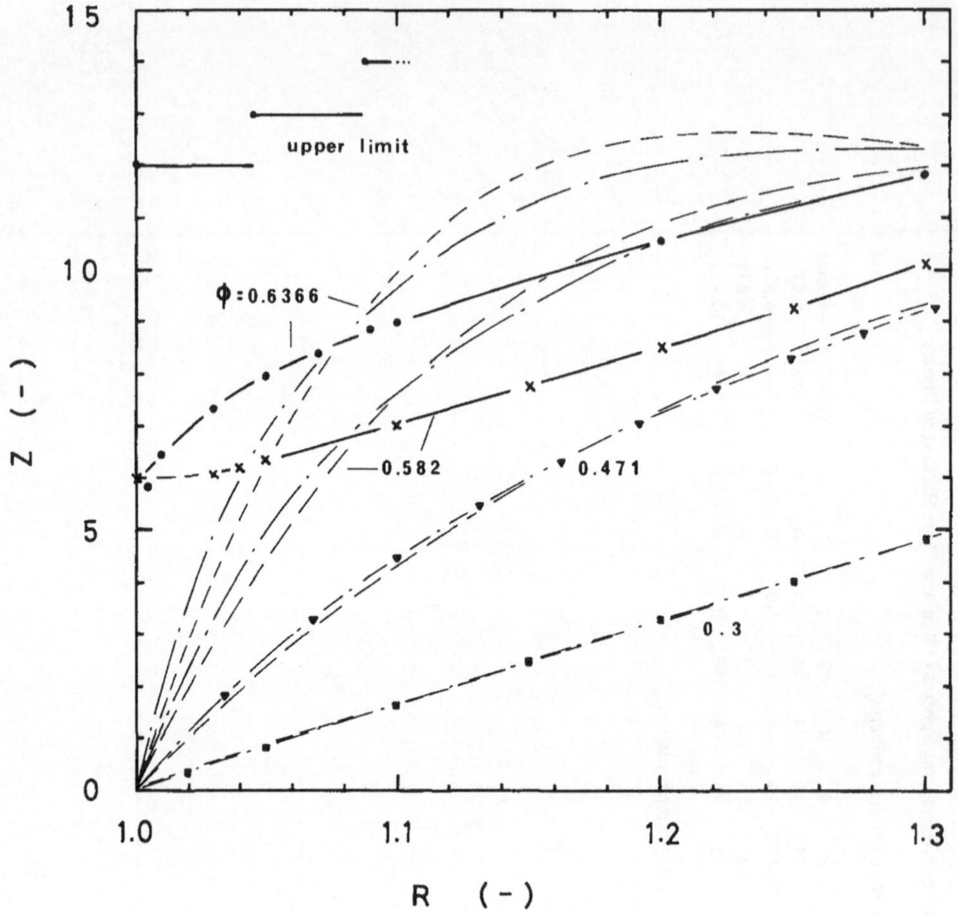

FIGURE 3. Relation between average cumulative number of neighbors Z and radial distance R. - - - and - • - , use of g (R) of Percus-Yevick[24] and its modification by Henderson & Grundke,[18] respectively.

Figure 3 is a plot of the average cumulative number of neighbors at distance R from a central sphere. The data for random dispersion are also depicted. This figure shows the difference in the structures of the packed and dispersed systems in the limit of $R \rightarrow 1.0$. Theoretical curves obtained from integrating the radial distribution function agree with the computer experiments for the random dispersion, but not for the random packing. The stairwise lines in Figure 3 depict the upper limit of this relation, which corresponds to the maximum number of the spheres that can be arranged on a spherical surface.[19] The coordination number measured by using japan paint[4] is 8.8 for the random close packing. In Figure 3 this corresponds to the data point at $R = 1.09$. In other words, the average cumulative number of neighbors at $R = 1.09$ corresponds to the coordination number of the random close packing. We do not know the rigorous definition of the coordination number.

From consideration of a random allocation of spheres into space cells, we can derive the probability distribution function of the local-mean volume fraction of the particles.[9,11] We can also obtain a model of the random packing as a random mixture of two regular packings such that the mean value and the variance remain unchanged. We eventually obtain a theoretical expression for the coordination number (Equation 5).[9,11]

$$Z = \frac{36}{\pi}\phi \quad : \quad 0.47 \gtrsim \phi \tag{5a}$$

$$= 20.7\phi - 4.35 \quad : \quad 0.47 \lesssim \phi \lesssim 0.7 \tag{5b}$$

As explained later, this relation gives agreement with experiments not only for sphere packings but also for the liquid state.

The structural study of the random assemblage of equal spheres has been summarized. Despite the fact that we have discussed only the sphere assemblages, lots of problems still remain unsolved. The case for nonspherical particles is beyond our present knowledge.

III. MOLECULAR PACKING SITUATION IN LIQUIDS

What is mentioned so far are the results obtained from the random assemblage of equal spheres. We next consider the molecular configuration of simple liquids. According to Bernal's idea,[5] liquids can be characterized as a homogeneous, coherent, and essentially irregular assemblage of particles. Liquid structure measured by X-ray or neutron diffraction analysis is the time average due to the self-diffusional motion of molecules. Alternatively the random packing of particles changes its internal structure from place to place. The space-average structure is considered to be equal to the time-average structure of liquids. This assumption of ergodicity might be a more detailed version of Bernal's idea.

Since we consider liquid as a random packing of molecules, the position of the first peak in the radial distribution curve corresponds to the molecular diameter d. Hence we can calculate the molecular packing density as the expression:

$$\phi = \frac{\pi}{6} d^3 \rho / m = \frac{\pi}{6} d^3 N/V = \frac{\pi}{6} \sqrt{2} / \left(\frac{V}{N\sigma^3}\right), \quad d = 2^{\frac{1}{6}}\sigma \tag{6}$$

where ϱ is the liquid density and m is the molecular mass. Alternatively we can obtain the coordination number of the system from integrating the area of the first peak in the radial distribution curve. We thus obtain a relation between the coordination number and the molecular packing density.

Equation 5 can apply not only for powder systems but also for molecular systems. Under the coherence assumption of molecules, the Lennard-Jones (6,12) potential equation yields the molecular diameter of $d = 2^{1/6}\sigma$. Namely, we consider liquid as a random packing of particles having the diameter d, each pair of the molecules being subjected to the potential energy $(-\varepsilon)$. Although being valid only for spherical, nonpolar molecules such as argon and krypton, we use the Lennard-Jones potential equation for all liquids by modifying it with the molecular shape parameter.

The classical partition function for a system of N molecules is given by Equation 7:

$$f_N = \left(\frac{2\pi m k T}{h^2}\right)^{\frac{3}{2}N} v_f^N \exp\left[-\frac{N\psi}{2kT}\right] \tag{7}$$

where k is Boltzmann's constant, T is the absolute temperature, h is Planck's constant, v_f is the the free volume of a single molecule, and ψ is the potential energy between one molecule and all other molecules. In the gaseous state, the free volume of a single molecule can be expressed by Equation 8,

$$v_f = \frac{eV_f}{N} \quad \text{for gases} \tag{8}$$

where V_f is the empty volume of the system.

As the density of the tightest possible packing of spheres is $\pi\sqrt{2}/6$, we have assumed Equation 9 for V_f:

$$V_f = V\left[1 - \frac{\phi}{\left(\frac{\pi}{6}\sqrt{2}/\lambda\right)}\right] \tag{9}$$

where V is the total volume, ϕ is the bulk- mean particle volume fraction, and λ is the molecular shape parameter which adjusts the closest packing density for the case of the nonspherical particles.

As mentioned before, it follows from the coherency assumption of particles that the molecular diameter d corresponds to the position of the first peak in the radial distribution curve. If we assume the Lennard-Jones (6,12) pair potential, d = $2^{1/6}\omega$ and the bulk-mean particle volume fraction becomes as Equation 6. Here we have denoted the potential parameters by ε and σ. Similarly, from the coherency assumption we obtain the potential energy of interaction, Equation 10:

$$\psi = -Z\epsilon \tag{10}$$

where Z is the coordination number given by Equation 5. The interaction with more distant molecules is disregarded. Therefore, using the thermodynamic relation for pressure, we obtain the equation of state, Equation 11:

$$\frac{pV}{NkT} = \frac{1}{1 - \frac{\lambda\phi}{\frac{\pi}{6}\sqrt{2}}} - \frac{18\phi}{\pi\left(\frac{kT}{\epsilon}\right)} \tag{11}$$

where ϕ <; 0.47. This equation is the same in form as famous van der Waal's equation. Moreover, all parameters in Equation 11 have explicit meaning. Equation 11 gives the critical point properties in Equation 12:

$$\lambda\frac{kT}{\epsilon} = \frac{8}{9}\sqrt{2}\,,\ \lambda\phi_c = \lambda\frac{\frac{\pi}{6}\sqrt{2}}{\left(\frac{V_c}{N\sigma^3}\right)} = \frac{\frac{\pi}{6}\sqrt{2}}{3} \tag{12}$$

We can obtain the potential parameters ε and σ from the critical temperature and volume using the expression 12. In practical calculation the numerical value 3.0 in the right-hand side of the expression 12 should be replaced by the experimental value 3.15, which yields the packing density $\lambda\phi_c = 0.235$ at the critical point. For the molecular shape parameter we can use the semiempirical relation in Equation 13:[14,20]

$$\lambda = \frac{1}{0.59} - \left(\frac{1}{0.59} - 1\right) \times 10^{-1.5w} \tag{13}$$

where w in the exponent is Pitzer's acentric factor which expresses the effect of the noncentral forces and can be calculated from the vapor-liquid equilibrium data.

In the liquid state we consider a cell model for the free volume v_f and use Equation 5b for the coordination number. We eventually lead to the equation of state of liquids, Equation 14.

$$\frac{pV}{NkT} = \frac{0.59\lambda}{1 - \left(\dfrac{\lambda\phi}{\frac{\pi}{6}\sqrt{2}}\right)^{1/3}} - \frac{20.7 \times 0.858\lambda\phi}{2\left(\dfrac{kT}{\epsilon}\right)} \tag{14}$$

This is the same in form as that of Renon et al.[20] which agrees fairly well with various thermodynamic properties for a variety of liquids. With the assumption that the effect of pressure is negligible, i.e., $p = 0$, Equation 14 gives a relation between $\lambda\phi$ vs. kT/ε, and we obtain Equation 15 in overall for a variety of liquids, including liquid metals in the case of the melting point.

$$\lambda\phi_b = 0.55 \quad \text{for boiling point;}$$

$$\lambda\phi_m = 0.64 \quad \text{for melting point} \tag{15}$$

For spherical molecules, $\lambda = 1$ and $\phi_b = 0.55$ corresponds to the random loose packing of equal spheres; $\phi_m = 0.64$ corresponds to the random close packing. In this fashion, we have obtained a general comparison between random structures of molecular systems and powder assemblages.

By the Lennard-Jones potential parameter obtained from the critical and melting temperatures, we can classify an ordinary state of pure substances: gas for $\varepsilon/k < 500$, liquid for $500 \lesssim \varepsilon/k \lesssim 800°K$, and solid (metal) for $\varepsilon/k > 800°$ K. It is one of the interesting results from this structural study that water at ordinary temperature is a liquid of looser packing density ($\phi \approx 0.42$) with stronger molecular interactions ($\varepsilon/k = 762°$ K), but mercury is a liquid of closer packing density ($\phi = 0.64$) with weaker atomic interacions ($\varepsilon/k = 569°$ K). Water especially is very porous so that it can contain various substances in its empty space. However, the pore sizes of water are smaller than any other molecules and, moreover, the interaction among molecules is very strong. Therefore it can be said that water is unlikely to dissolve any other molecules without any mechanical disturbances such as a mixing operation.

The structural study can also deal with the gas solubilities in liquids. The gaseous solute is considered to move freely through the pores in the liquid solvent, and we adopt the idea of the empty volume for the free volume v_f, (Equation 16)

$$v_f = \frac{V_e}{N_A + N_B}\left[1 - \frac{\phi_m}{\frac{\pi}{6}\sqrt{2}/\lambda_B}\right] \quad : \text{ empty volume theory} \tag{16}$$

where V_e is the volume of the liquid phase, N is the number of the molecules in the liquid phase, the subscripts A and B denote components A and B, respectively, and ϕ_m is the molecular packing density of the liquid expressed by Equation 17.

$$\phi_m = \frac{\frac{\pi}{6}\sqrt{2}\,\sigma_A^3 N_A + \frac{\pi}{6}\sqrt{2}\,\sigma_B^3 N_B}{V_e} \tag{17}$$

Since we are dealing with a very dilute solution, the closest packing density of the B component is considered in Equation 16. This empty-volume theory can apply for most solvents. Alternatively, liquid metals and water at ordinary temperatures require another treatment. Since the pores in the instantaneous packing structure of water molecules are not large enough for solute gas molecules to penetrate, the gases are not likely to dissolve: once dissolved, they may be trapped firmly by the water molecules so that they cannot escape from the cage easily. In other words, the movement of the gas molecule is restricted by those of the surrounding water molecules. Thus we can assume that the free volume of the gas molecule is equal to that of the water molecules, yielding Equation 18.

$$v_f = \frac{\pi}{6}(d_{cell}) - d_B)^3 \quad : \text{ cell theory} \tag{18}$$

$$\frac{\pi}{6} d^3_{cell} = \frac{\frac{\pi}{6}\sqrt{2}\, V_e}{N_A + N_B} \tag{19}$$

This equation is based upon the cell theory. Since the spherical cells do not fill the whole volume of liquid phase, V_e is modified with the closest packing density. In this manner, we consider a dilute binary mixture of solute gas A and solvent liquid B. Using the free volume mentioned above and the coordination number, we can derive the partial pressure of solute, whose differentiation with respect to the solute concentration in the liquid phase eventually yields the Henry's law constant at infinite dilution.[16]

IV. CONCLUDING REMARKS

We have discussed a comparison between random structures of molecular systems and powder assemblages. If we have detailed knowledge about structures of the random systems, we can explain a variety of physicochemical properties of matter. Powder assemblages can play important roles as a reference system. In this text, we have excluded the assemblage of nonspherical particles, simply because we do not have enough information about it. Even in the case of the spherical particles, lots of problems still remain unsolved. In order to deal with the structural study experimentally, we need to develop a computerized technique for measuring two- and three-dimensional structures.

V. ACKNOWLEDGEMENT

I am grateful to Professor T. P. Meloy for his encouragement, Professor J. K. Beddow, and the National Science Foundation, for having afforded me this opportunity. I was on leave from the Department of Chemical Process Engineering, Hokkaido University, Sapporo, Japan. I thank Professor E. M. Tory for my visiting research in Canada.

REFERENCES

1. **Bansal, P. P. and Ardell, A. J.,** Average nearest-neighbor distances between uniformly distributed finite particles, *Metallography*, 5, 97, 1972.
2. **Bannett, C. H.,** Serially deposited amorphous aggregates of hard spheres, *J. Appl. Phys.*, 43, 2727, 1972.
3. **Bernal, J. D.,** A geometrical approach to the structure of liquids, *Nature (London)*, 183, 141, 1959.
4. **Bernal, J. D. and Mason, J.,** Coordination of randomly packed spheres, *Nature (London)*, 188, 910, 1960.
5. **Bernal, J. D.,** The Bakerian lecture, 1962, the structure of liquids, *Proc. Roy. Soc. (London)*, A280, 299, 1964.
6. **Bernal, J. D., Cherry, I. A., Finney, J. L., and Knight, K. R.,** *Br. J. Phys. E.*, 3, 388, 1970.
7. **Chandrasekhar, S.,** *Rev. Mod. Phys.*, 15, 1, 1943.
8. **Finney, J., C.,** Random packings and the structure of simple liquids. I. The geometry of random close packing, *Proc. Roy. Soc. (London)*, A319, 479, 1970.
9. **Gotoh, K.,** Spatial distribution of equal-sized particles in a random assemblage and minimum size of a representative sample, *Ind. Eng. Chem. Fundam.*, 10, 161, 1971.
10. **Gotoh, K.,** Distribution of particle spacings in a random dispersion, *Ind. Eng. Chem. Fundam.*, 10, 315, 1971.
11. **Gotoh, K.,** Liquid structure and the co-ordination of equal spheres in random assemblage, *Nature (London) Phys. Sci.*, 231, 108, 1971.
12. **Gotoh, K.,** Molecular configuration of spherical, non-polar, dense gases and liquids, *Nature (London) Phys. Sci.*, 232, 64, 1971.
13. **Gotoh, K.,** Equation of state of simple liquids from coordination number in random assemblage, *Nature (London) Phys. Sci.*, 234, 138, 1971.
14. **Gotoh, K.,** Liquid structure and Lennard-Jones 6,12 pair potential parameters, *Nature (London) Phys. Sci.*, 239, 154, 1971.
15. **Gotoh, K. and Finney, J. L.,** Statistical geometrical approach to random packing density of equal spheres, *Nature (London)*, 252, 202, 1974.
16. **Gotoh, K.,** Solubilities of nonreacting gases in liquids from the free-volume theory, *Ind. Eng. Chem...*
17. **Haughey, D. P. and Beveridge, G. S. G.,**Local voidage variation in a randomly packed bed of equal-sized spheres, *Chem. Eng. Sci.*, 21, 905, 1966.
18. **Henderson, D. and Grundke, E. W.,** *J. Chem. Phys.*, 63, 601, 1975.
19. **Mackay, A. L. Finney, J. L., and Gotoh, K.,** The closest packing of equal spheres on a spherical surface, *Acta Crystallogr.*, A33, 98, 1977.
20. **Renon, H., Eckert, C. A., and Prausnitz, J. M.,** Molecular thermodynamics of simple liquids, *Ind. Eng. Chem. Fundam.*, 6, 52, 1967.
21. **Scott, G. D.,** Packing of equal spheres, *Nature (London)*, 188, 908, 1960.
22. **Scott, G. D.,** Radial distribution of the random close packing of equal spheres, *Nature (London)*, 194, 956, 1962.

23. **Scott, G. D. and Kilgour, D. M.,** The density of random close packing of spheres, *Br. J. Appl. Phys.*, 2 (Series 2), 863, 1969.
24. **Smith, W. R. and Henderson, D.,** Analytical representation of the Percus-Yevick hard-sphere radial distribution function, *Mol. Phys.*, 19, 411, 1970.
25. **Tory, E. M., Church, B. H., Tam, M. K., and Ratner, M.,** Simulated random packing of equal spheres, *Can. J. Chem. Eng.*, 51, 484, 1973.
26. **Visscher, W. M. and Bolsterli, M.,** Random packing of equal and unequal spheres in two and three dimensions, *Nature (London)*, 239, 504, 1972.

INDEX

O